East Asian experience in environmental governance

East Asian experience in environmental governance: Response in a rapidly developing region

Edited by Zafar Adeel

TOKYO • NEW YORK • PARIS

United Nations University Press
The United Nations University, 53-70, Jingumae 5-chome,
Shibuya-ku, Tokyo, 150-8925, Japan
Tel: +81-3-3499-2811 Fax: +81-3-3406-7345
E-mail: sales@hq.unu.edu
General enquiries: press@hq.unu.edu
www.unu.edu

United Nations University Office in North America
2 United Nations Plaza, Room DC2-2062, New York,
NY 10017, USA
Tel: +1-212-963-6387 Fax: +1-212-371-9454
E-mail: unuona@ony.unu.edu

United Nations University Press is the publishing division of the United Nations University.

Cover design by Joyce C. Weston
Cover photograph "Shanghai Skyline and Huangpujiang River" © 2000 Guang Hui China Tourism Press/The Image Bank

Printed in the United States of America

UNUP-1072
ISBN 92-808-1072-3

Library of Congress Cataloging-in-Publication Data

East Asian experience in environmental governance: response in a rapidly developing region / edited by Zafar Adeel.
p. cm.
Includes bibliographical references and index.
ISBN 92-808-1072-3
1. Environmental policy—East Asia—Case studies. 2. Environmental policy—Malaysia. 3. Environmental policy—Thailand. I. Adeel, Zafar.
GE190.E18 E17 2003
363.7'0095—dc21 2003006855

Contents

List of tables and figures

Tables

Figures

Preface

The United Nations University (UNU) has always considered the East Asian region as a priority area for its activities. This emphasis is partly driven by the UNU's location in Japan, but it is also based on the consideration that most nations in this region are developing countries. With the UNU's mandate to build networks of researchers and scholars and to develop the capacity of individuals and institutions to undertake research, the focus is always on developing countries. Environmental governance in this region has also received due attention from the UNU; in fact, the UNU has since 1996 undertaken a programme both to monitor the environmental quality in the region and to outline prescriptions for environmental policies. This programme has been possible largely due to diligent collaboration of a network of researchers, professionals, and scholars working in the region. This book presents a compilation of papers on environmental governance contributed by the members of this network.

In the context of this book, it is important to start with defining the geographical extent of "East Asia" as it is used here, because a number of delineations are available in accordance with various geographic and economic emphasis. For the purposes of this book, East Asia is defined as a region extending from Japan and China in the east to Thailand in the west and down to Indonesia and the Philippines. In other settings, the combination of Viet Nam, Laos, Cambodia, Thailand, Malaysia, Singapore, and Indonesia is considered as part of South-East Asia (such as in ASEAN) or simply South Asia. In still different settings, the same area may be referred to as the Asia Pacific region with the inclusion of the Pacific island states (such as within the framework of the Asian Productivity Organization); the World Bank terms this same region as "East Asia and the Pacific". In recent terminology used by the UN Environment Programme the whole region has been divided into three segments: South-East Asia, comprising Brunei Darussalam, Indonesia, Malaysia, the Philippines, and Singapore; the Greater Mekong region, comprising Cambodia, China (Yunnan province), Laos, Myanmar, Thailand, and Viet Nam; and North-West Pacific and East Asia, comprising China, the Republic of Korea, Japan, and Mongolia. A number of other definitions are also used.

Suffice to say that the multitude of choices for compartmentalization into regions and subregions frequently reflects the institutional context. Often, the focus on lumping together countries for administrative convenience may produce counter-intuitive groupings. For the purposes of this volume, it may be more important to step back from the semantics of regional definition and focus on similarities and differences among countries in this region in the context of environmental governance. This book therefore adopts a somewhat simplified regional definition of East Asia, with due recognition of the other existing definitions. Five countries have

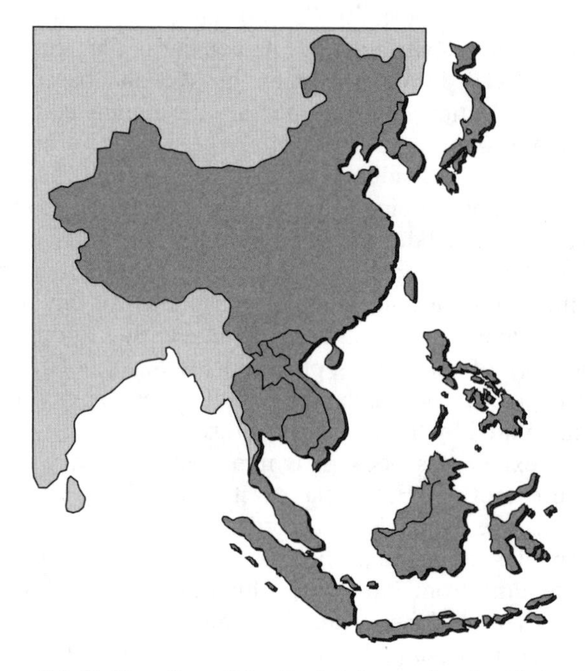

Figure P.1 Delineation of the East Asian region in this book

been selected as representatives of the region: China, Japan, the Republic of Korea, Malaysia, and Thailand. This selection is deliberately made so that highly industrialized, industrializing, and developing economies are adequately represented. The grouping also provides a mix of political and historical backgrounds that are diverse enough to provide a glimpse of the "typical" East Asian governance mechanisms. The region has seen considerable growth in its economy, industrial base, and population. Interestingly, all three of these factors are often linked to degradation of environmental resources. The areas selected for deeper analysis in the book – pesticide management, water quality and resources, and air pollution management – cut across all three of the above factors and also include agriculture as a distinct element of environmental governance.

Overall, this volume provides a broad-brush overview of the existing environmental governance regime in the East Asian region. It is anticipated that the findings from the book, and the case studies contained herein, can lead to a fundamental understanding of what works and what does not in this region; such conclusions can often be enlightening for researchers and policy-makers. Ultimately, effective and meaningful environmental governance can ensure long-term sustainability of the remarkable industrial and economic growth observed in this region. Even more importantly, it would ensure that our children and grandchildren can inherit a region that is prosperous yet rich in culture, environmental resources, and natural beauty.

Zafar Adeel
April 2002, Tokyo

1

Introduction to environmental governance concepts in the East Asian context

Zafar Adeel

The East Asian miracle

East Asia has emerged as a dynamic and rapidly evolving region during the past three decades, like a phoenix rising from the ashes of colonialism. The East Asian "experience" has elicited a great deal of interest from observers and scholars. This experience ostensibly pertains to the economic and industrial growth that has brought countries in the region to the doorstep of the first world. The World Bank, for example, indicates that most of the so-called miracle economies are concentrated in this region. These economies went through a major crisis starting in late 1997, the lingering effects of which lasted through the year 2000. Nevertheless, because there are already signs of a strong economic recovery and the environmental governance regime is marginally affected, it is worthwhile to understand the underpinnings of this economic miracle and what role it has played in the development of the environmental governance regime.

Since the 1970s the industrial growth in this region has raised justifiable concerns among environmentalists as to its adverse impacts on the bountiful natural resources and diverse ecosystems. Considerable environmental degradation is apparent on a regionwide basis, ranging from deforestation and loss of biodiversity to pollution of waterways and fresh water resources. Environmentalists, with the help of activist groups and non-governmental organizations (NGOs), have played a key role in bringing to light these impacts while stirring agitation against most major development initiatives. A limited discussion of the role of these groups is provided in the final section of this book.

The response of the region through the development of an environmental governance regime can be viewed as a complex process. This process is more complicated than either environmental activists or the governments in the region will readily admit. To understand better the East Asian experience and the nascent environmental governance regime, it is quite important to know the political and historical context of the region. This context, in fact, drives how any governance regime is perceived and developed by the countries and peoples of the region. At the same time, the historical evaluation helps us understand the current situation in the region. Islam and Chowdhury (1997) provide a much more in-depth analysis of this background.

Political and historical context

The emergence of an East Asian bloc in the global economy can be traced back to a number of "historical accidents" (Petrie 1994). Three main categories of these historical accidents can be easily identified in a chronological sequence:

- Western imperialism in the region, particularly the introduction of the treaty port system;
- Japanese imperialist endeavours and the role they played in economic development of the region;
- influence exercised by the USA in the wake of post-Second World War geo-politics in the Pacific region.

During the middle of the nineteenth century, a unique cooperative imperialism emerged between the European powers, notably the British, Russian, Prussian, Portuguese, Danish, Dutch, and Spanish empires, as well as the USA (Petrie 1994). This evolved into a treaty port system which resulted in a significant portion of global trade being channelled through this region. The most salient examples of this were the ports of Hong Kong, Manila, Singapore, and Shanghai. This system provided the basis for the involvement and engagement of the region in international trade.

The influence of Japanese imperialism on economies in this region leading up to the Second World War is relatively well documented (Ho 1984). This economic influence in Taiwan, Korea, and China came about in the form of massive investments in selected sectors such as transport and communications, education, and manufacturing. The vested interest in such investments was to exploit fully any complementarities to the benefit of Japanese industries. This led to two major consequences for the region (Petrie 1994). First, Japan emerged as a key player in the regional economy and its trade in the region surged by about 70 per cent prior to the Second World War. Second, it helped imperial Japan in implementing its hegemonic designs through the Greater East Asia Coprosperity Sphere (GEACS).

The Second World War brought some abrupt changes to this set-up. The defeat of Japan in the war led to the emergence of the USA as a major political force in the region. The ensuing period has to be viewed in the context of geo-politics from the 1950s through the 1970s. Particularly, the strategic alignment of the USA during the Cold War and its involvement in the conflicts on the Korean peninsula and in Viet Nam were central to its role in this region (Haggard 1988). There was a massive influx of American aid to the Republic of Korea, Taiwan, and the Philippines, and an overall reorientation of the regional trade towards the USA (Islam and Chowdhury 1997). This economic and political orientation provided the platform from which the so-called miracle economies were launched in the region during the mid- to late 1970s.

This discussion provides a brief backdrop against which the recent industrial growth and economic boom (and bust) can be evaluated. It lends some weight to regional interpretation of the development in this region, in whatever manner it may be delineated by various international organizations. It also provides a hint of the colonial imprint on the governance structures in the region. The most salient feature of this colonial imprint is the depletion of natural resources (Parnwell and Bryant 1996). A number of natural resources were, and in some cases still are, being exploited beyond any hope of recovery. Some prominent examples of this exploitation during the colonial era include rice, teak, and minerals from Thailand, coffee and sugar from Java (Indonesia), and tin, palm oil, and rubber from what is now Malaysia.

The introduction of political independence in this region and the end of colonial rule has influenced the social, political, and governmental structures. For example, a major common feature has been alignment of élite economic and political groups for mutual interest. These influences do not follow a uniform pattern across the board. It is therefore instructive to take a closer look at the political and economic set-up of the five countries included in this book.

China has developed itself into a major player in the regional as well as global economy and politics. Although the Chinese economy still falls under the umbrella of "developing nations", it is the world's seventh largest economy based on GNP and second largest when GNP is converted to purchasing power parity (World Bank 2000). The most salient feature of this economy has been a sustained double-digit growth through the early 1990s. This is in great part attributable to the dynamic economic reforms introduced since 1993, which are designed to evolve China into a prosperous nation. In addition to financial and fiscal reforms, a broad range of reforms has also been put into action within legal and social welfare systems. Politically, these reforms are also remarkable considering the political set-up whereby the Chinese Communist Party enforces control over all the governmental institutions at every level. It is important to note here that the incredible economic and industrial growth has not been without its environmental repercussions. Some examples of these adverse impacts on the water quality in China are highlighted in Chapter 3 of this book.

Currently, Japan is the most significant player in the regional economy. This is in direct correlation with its importance in the global economy, being the second largest in terms of GDP per capita and third largest in terms of exports of goods and services (World Bank 2000). A majority of Japan's trade with developing countries is focused towards the East Asian region. Understandably, Japan has also led the regional efforts to create a formal economic bloc under the Asia Pacific Economic Cooperation (APEC) umbrella. Internally, Japan has followed a very strict path towards industrial development and economic growth since the Second World War. Particularly, industrial growth during the 1960s and 1970s led to the "bubble economy" of the 1980s. Frequently this growth led to environmental degradation and natural resources depletion, often in other East Asian countries. When compared with other developed nations the emergence of an environmental governance framework in Japan, as presented in Chapter 8, has progressed at a somewhat slower pace.

This book limits its evaluation of the situation on the Korean peninsula to the Republic of Korea (simply referred to as Korea hereafter). Historically, Korea has a colonial imprint of Japanese imperialism, particularly from the late nineteenth century to the end of the Second World War. However, partition of the country into north and south zones following the war has also had significant long-term implications. As mentioned earlier, geo-political considerations and the American influence have greatly impacted on economic growth patterns in Korea. The country has gone through a series of political perturbations, culminating in the adoption of a new constitution in 1987. However, the economic development initiated in 1962 led to tremendous growth of the industrial sector, such that by the end of the 1980s industrial output accounted for 43 per cent of the GDP (Islam and Chowdhury 1997). Since the late 1980s, the country has also focused on developing legislation that can form part of the environmental governance regime. This is discussed in the context of air quality management in Chapter 7 of this book.

Malaysia now has a parliamentary democracy that represents an ethnically diverse society comprising Malay, Chinese, Indian, and other cultures. From the last quarter of the nineteenth century till 1957, Malaysia was almost continually under direct British influence and rule; the exception being the brief Japanese occupation during the 1942–1945 period. A systematic approach to economic planning in the form of the "new economic policy" in 1971 and the "new development policy" in 1991 has been quite successful in overcoming any racial barriers to development in a multi-ethnic society. As a result, the Malaysian economy has seen significant growth – particularly posting an annual growth of over 8 per cent in the late 1980s and through the 1990s. This has helped bring the country inside the OECD envelope, and it is anticipated that it will enjoy developed country status by 2020. However, the results of this growth for the environment have not been altogether favourable. As described in Chapters 2 and 5, the Malaysian government has engaged in a proactive approach in development of environmental legislation and public awareness programmes.

Historically speaking, Thailand has a unique character in the group of five countries discussed in this book. It has never been colonized formally by either the European or Japanese empires, although it developed alliances with both through various stages of its recent history. Its politics was dominated by military influence until 1992, when liberal democracy was fully established. Regional conflicts and geo-politics have placed Thailand in a position to receive military and financial support from the USA (Dixon 1995). Such support has resulted in a boost to the Thai economy. Additionally, orientation of the Thai economy towards export-oriented manufacturing helped Thailand to reach double figures in annual economic growth during the early 1990s (Islam and Chowdhury 1997). Despite the economic slump in 1997–1998, the national economy has shown a strong recovery. Development of environmental institutions, as described in Chapters 4 and 6, is still progressing towards maturity.

Factors related to environmental impacts

Economic growth

The economic crisis in Asia, starting in 1997 and reaching full bloom in 1998, has triggered an extensive debate on the so-called myth of the Asian miracle. The East Asian economies had been praised, equally extensively, for their rapid growth and model development strategies prior to the crisis (Weder 1999). However, the perspective of environmental governance, as discussed in this book, is somewhat different to that held by most economists. Seen from this view, the economic growth in East Asia was coupled to rapid growth of both industrial and urban sectors,

Table 1.1 Growth rate of GDP, percent per annum

Country[a]	1981–90	1991	1992	1993	1994	1995	1996	1997	1998	1999	2000	2001
China	10.4	9.3	14.2	13.5	11.8	10.2	9.6	8.8	7.8	7.1	8.0	7.3
Korea	12.7	9.1	5.1	5.8	8.4	9.2	6.8	5.0	−6.7	10.9	9.3	3.0
Malaysia	5.2	8.6	7.8	8.3	9.2	9.3	10.0	7.3	−7.4	6.1	8.3	0.4
Thailand	7.9	8.5	8.1	8.3	8.7	8.6	5.9	−1.4	−10.8	4.4	4.6	1.8
Japan[b]	4.0	3.8	1.0	0.3	0.6	1.5	5.0	1.4	−2.8	1.0	1.5	

a. Data for 1981–1995 from ADB (1997); Data for 1996–2001 from ADB (2002)
b. Data for Japan from IMF (1999)

which in turn led to a certain set of environmental impacts. These impacts on natural systems are persistent and long term, and possess an "inertia" of their own. Although the economies in the region slowed down during 1998 and gradually recovered during 1999 and 2000, the environmental impacts did not follow a similar fluctuating trend and were more or less persistent. For example, the levels of pollution in coastal waters show no incremental reduction during the 1997–2000 period.

Economic growth in the region was strongly tied to export-oriented policies, high savings rates, sound macroeconomic policies, and strong institutional frameworks (Weder 1999). As already noted for the five selected countries, the rule of law, political stability, and historical context play an important role in defining a country's growth. This experience, despite some variability from country to country, has certain commonalities, including colonial rule – or at least heavy influence from Europe or the USA – political instability, correlation to regional and global trade, and geo-political influences. Against this backdrop, one may compare the economic performance of the five countries. Table 1.1 provides an overview of GDP growth rates from the 1970s through to the present time. A gross similarity between the economies, not including Japan, is their sustained growth at levels above 5 per cent, with China leading the race with double-digit GDP growth figures.

Explosive population growth

As can be observed from Table 1.2, all countries in the East Asian region have been successful in reducing population growth rates. Nevertheless, the population growth in this region is still quite significant, with the exception of Japan where population growth has been steadily decelerating. This is important when seen in terms of absolute numbers; the five countries listed here comprise a quarter of the global population. On a

Table 1.2 Population statistics for East Asia

Country	Average annual growth rate (%)		Total population (millions)		% living in urban areas
	1980–1990	1990–1998	1980	1999	1999
China	2.3	1.2	981	1,250	32
Japan	0.8	0.3	117	127	79
Korea	2.0	1.1	38	47	81
Malaysia	4.8	2.8	14	23	57
Thailand	2.7	1.4	47	62	21

Source: World Bank (2000) and World Bank (2001)

regional scale, the population has more than doubled during the latter half of the twentieth century (UN ESCAP 2000). This growth in population has been offset by a much faster growth in economies, as observed earlier. However, the other side of this coin shows a trend of unsustainable consumption of natural resources.

Even more important than the gross increase in numbers is the trend of increasing urban populations which far outpaces the overall growth rates. This is, in part, due to migration of people to urban areas in greater numbers. As shown in Table 1.3, a number of the fastest-growing existing and potential mega-cities – those with population in excess of 10 million – are located in this region. Of the five countries considered here, only China and Thailand have a majority of their population still living in rural areas; Japan, Korea, and Malaysia have a vast majority of their populace living in urban settings. The growth of population, particularly

Table 1.3 Top 10 fastest-growing mega-cities in Asia

City and country	% annual growth
Dhaka, Bangladesh	4.0
Karachi, Pakistan	3.3
Istanbul, Turkey	3.1
Bombay, India	2.7
Delhi, India	2.6
Jakarta, Indonesia	2.3
Manila, Philippines	2.2
Tehran, Iran	2.0
Calcutta, India	1.8
Beijing, China	1.5
Shanghai, China	1.3
Tianjin, China	1.2

Source: ADB (1996)

urban population, has serious consequences for utilization of natural resources and impacts on the environmental compartments. In some of the countries in this region, the absence of an adequate urban infrastructure is the root cause of these environmental impacts (UN ESCAP 2000).

Major environmental impacts

The East Asian miracle has left a broad range of environmental impacts in its wake. It is indeed challenging to come up with a complete list of these, but below are some of the greatest impacts in this region:

- degradation and destructions of forests;
- massive soil erosion and desertification, partly due to improper land use and management practices;
- pollution of air by particulates (aerosols) as well as greenhouse gases;
- contamination of fresh water resources with urban, industrial, and agricultural wastes;
- accumulation of pollutants in coastal areas and impacts on fisheries;
- transport air pollution in the form of smoke and haze on a regional scale.

Each of these impacts is present to varying degrees and magnitudes in the five countries discussed in this book. By nature, these problems are not limited by national boundaries and thus require regional responses. It is also important to consider the correlation of these environmental issues with broader socio-economic factors. Most importantly, degradation of the environment has serious consequences for food security, poverty, quality of life, and economic stability.

Environmental governance in East Asia

Governance and environmental governance

It is somewhat difficult to grasp the concept of governance and even more so to define environmental governance. Governance can be defined as a complex set of values, norms, processes, and institutions by which society manages its development and resolves conflict, formally or informally. Generally speaking, governance involves state players (various levels of government), but also civil society at the local, national, regional, and global levels (Hempel 1996). It also involves political institutions and sets of rules, including decision-making procedures that give rise to social practices. Historically, such governance has been an essential component of global security and economic development issues. The addition of environmental issues to this list is relatively new and has emerged over the past two or three decades. The realization that most environmental problems and concerns are of a transboundary nature and that we all share common natural resources has highlighted the need for an environmental governance regime that is global in its extent.

Environmental governance comprises complex governance elements dealing with various environmental compartments in an integrated manner through involvement of a wide range of actors and stakeholders. Environmental governance is also applied at a range of scales, from local to national to global. On the whole, it is the key for achieving balance between consumption, human welfare, and environmental consequences. However, it is based on a relatively short history and relies upon still-emerging scientific disciplines.

Implementation of environmental governance

Development of environmental governance regimes and the related institutions in the East Asian region has clearly been affected by the UN Conference on Environment and Development (UNCED 1992) held in Rio, and Agenda 21. A number of approaches for establishing regulatory and institutional frameworks to achieve sustainable development were identified in Agenda 21. These have been recognized by the governments in the region and implemented with varying degrees of success. It is also interesting to note that promoting approaches to sustainable development has had a side-effect of encouraging participatory approaches, where NGOs and the private sector have a greater role to play. The governmental structures pertaining to environment have also seen a gradual streamlining, as is apparent from the information presented in Table 1.4.

A recent state of the environment report prepared by UN ESCAP (2000) highlights that there is increased promotion of community participation, decentralization of administration, and integration of national and local decision-making processes in the region. As can be observed from Table 1.4, there is apparent vertical integration at the governmental level that is helpful in implementation of effective environmental governance. Each country has a dedicated administrative facility at the ministerial level that is charged with coordination amongst related ministries and implementation of and compliance with governmental programmes.

Selection of sectors for analysis

Three sectors have been selected for a more detailed evaluation in this book: agriculture (with emphasis on pesticide usage), water resources

Table 1.4 National implementation of environmental governance

Country	Vision document	Policy institution	Executing agency	Apex national council
China	China's Agenda 21	State Environmental Protection Administration	State Environmental Protection Administration	
Korea	Green Vision 21	Ministry of Environment (MoE)	Various commissions/ bureaus under MoE	Environmental Conservation Committee
Japan	National Environment Plan	Environmental Agency	Various departments/ bureaus	Japan Council for Sustainable Development
Malaysia	Vision 2020	Ministry of Science, Technology and Environment	Department of Environment	Environmental Quality Council
Thailand	National Plan	Ministry of Science, Technology, and Environment (MoSTE)	Various departments under MoSTE	National Environmental Board

Source: Adapted from UN ESCAP (2000)

(and quality issues), and air pollution. A close scrutiny of each of these sectors is provided in the subsequent sections. This includes looking at the core issues and drivers behind them. The governance structures for each of these sectors are investigated to assess their strengths and weaknesses, as well as correlation to environmental governance at national scale and linkage to sustainable development. As a result of this assessment, some patterns and trends emerge. At the same time, the unique situation in each country leads to a disparity in approaches adopted. These are discussed to a greater extent in the final section of the book.

In all the five countries, the agricultural sector remains a predominant player in the national economy, making it an important component of the environmental governance regime. This sector has been under increasing pressure to meet the needs of the burgeoning populace while faced with reduction in arable lands. A most poignant example of meeting this challenge of food security can be seen in the case of rice production. A revolution in rice production started in Asia during the 1970s and continued through the 1980s. During this period tremendous achievements in rice production were made, in which an increase of about 50 per cent was recorded in the years from 1978 to 1989 in this region. This "green revolution" saved the region from serious food shortages during the 1970s when overall global food production was actually decreasing (Uitto 2000).

The driving force behind the green revolution was the technologies of chemical fertilizers for soil enrichment and pesticides and herbicides for crop management. Coupled with high-yielding rice varieties and increased land for rice production, remarkable progress was achieved. However, the price for this advanced technology and food security was paid by the environment. The use of chemical fertilizers, pesticides, and herbicides was critical for the success of the green revolution, but their indiscriminate use caused a long-term degradation of the environment (Hough 1998).

The environmental degradation is apparent at several levels. First, the pesticides used for agricultural purposes are persistent and resist natural degradation processes. As a result, they remain in the soils in agricultural areas and then leach away with water to rivers and coastal areas as well as underground aquifers. Second, this polluted water adversely impacts ecosystems that it comes in contact with. As an example, fish populations in rivers and coastal mangroves often have high levels of pesticides. Third, the food produced through these chemical technologies typically has residual levels of pesticides and insecticides. Lastly, the improper use of these technologies has been linked to accidental exposure of farmers to chemicals. This problem is particularly acute in the East Asian region, where the level of education is generally low and programmes for training pesticide users are not particularly successful.

The water resources in the East Asian region have served as the lifeline for economic and industrial growth. However, the growing demand for often limited water resources has pushed water on to the "endangered" list. At the same time, the existing water resources have also been polluted through run-off from industrial, urban, and agricultural activities. According to a recent study, the water resources in this region are projected (when using a conventional development scenario) to be at a "medium" level of vulnerability by the year 2025 (Raskin 1997). To avoid major problems in water availability, it is essential to establish programmes for integrated management of water resources and for protection of these resources from various pollution sources. It is also important to harness the financial and technical resources of the private sector for improving water supply and sanitation infrastructures. All of this requires well-defined national goals and governance mechanisms, which makes it an interesting study.

The air pollution problems in the East Asian countries have been primarily driven by two major factors: uninhibited industrial growth and urbanization. The development and implementation of air quality standards has lagged considerably behind those in the developed nations. This is partly reflected in inclusion of countries in this region in the Non-Annex I list of the Kyoto Protocol (under UN FCCC), with the exception of Japan. In layman's terms, the countries in this region are allowed to continue and/or increase emissions of greenhouse gases to achieve satisfactory growth in their industries and economies. Again, it is interesting to study the development of a governance regime to monitor and control air pollution in the region.

Overview

An equitable emphasis on the three selected sectors is intended by using case studies from the five selected countries. The authors of the case studies have linked the problems and issues to the governance structures in their respective countries. Often, the history of the development of these structures is also discussed, which provides insights into the shortcomings and limitations of the political processes involved. The role of various stakeholders, including government, the general public, NGOs, and industries, is described to complete the picture. The authors have also attempted to outline prescriptions for each sector in their respective country.

The book comprises four sections, with one section dedicated to each of the three sectors. The first section examines the management of pesticides in the agricultural sector of Malaysia (Abdullah and Sinnakkannu),

China (Hao and Yeru), and Thailand (Tabucanon). This sector is the most complex in terms of the number of players involved and the myriad of legislative enactments. It is interesting to observe the complex interrelationship between various laws and rules, while keeping in sight the limitations to their implementation on the ground. The second section focuses on water resource management in Malaysia and Thailand. Tabucanon, in her second contribution to this book, indicates that Thailand's perspective is driven by urban utilization of water and pollution issues. In contrast, Ahmad and Ali contend that Malaysia's water utilization patterns are largely driven by agricultural usage. While there are some similarities in the legislative framework of these two countries, inherently different approaches are adopted towards solving water management problems. The third section compares the air pollution issues and governance mechanisms in Korea (Lee and Adeel) and Japan (Yamauchi). The nature of the problems is somewhat similar, in part because of similar levels of industrial and economic development. The approaches to environmental governance are also somewhat similar in the two countries, with almost parallel development of environmental legislation.

The fourth section provides an overarching analysis of the governance structures in the region. An in-depth discussion of linkages of environmental protection and sustainable development to economic growth is undertaken. Paoletto and Termorshuizen outline a number of options for environmental governance through a comparison between approaches undertaken by the OECD countries, the USA, and the East Asian region. The final chapter (Adeel and Nakamoto) summarizes the findings of the earlier sections through a comparative evaluation. A synthesis of prescriptions for effective environmental governance is also provided.

REFERENCES

ADB. 1996. *Megacity Management in the Asian and Pacific Region*. Manila: Asian Development Bank.

Dixon, C. 1995. "Thailand – Economy", in *The Far East and Australasia*, 26. London: Europa Publications.

Haggard, S. 1988. "The politics of industrialization in the Republic of Korea and Taiwan", in Hughes, H. (ed.) *Achieving Industrialization in Asia*. Cambridge: Cambridge University Press.

Hempel, L. C. 1996. *Environmental Governance: The Global Challenge*. Washington, DC: Island Press.

Ho, S. P. S. 1984. "Colonialism and development: Korea, Taiwan and Kwantung", in Myers, R. H. and P. Petrie. (eds) *The Japanese Colonial Empire – 1895–1945*. Princeton: Princeton University Press.

Hough, P. 1998. *The Global Politics of Pesticides: Forging Consensus from Conflicting Interests*. London: Earthscan Publications.

Islam, I. and A. Chowdhury. 1997. *Asia-Pacific Economies: A Survey*. London: Routledge.

Raskin, P. 1997. *Comprehensive Assessment of the Freshwater Resources of the World*. Stockholm: Stockholm Environment Institute, pp. 66–67.

Parnwell, M. J. G. and R. L. Bryant (eds). 1996. *Environmental Change in South-East Asia: People, Politics and Sustainable Development*. London: Routledge.

Petrie, P. 1994. "The East Asian trading bloc: An analytical history", in Garnaut, R. and P. Drysdale. (eds) *Asia-Pacific Regionalism – Readings in International Economic Relations*. Sydney: HarperCollins.

Uitto, J. 2000. "Population, food and water in the 21st century", in Rose, J. (ed.) *Population Problems – Topical Issues*. New York: Routledge.

United Nations. 2001. *World Economic and Social Survey 2001 – Trends and Policies in the World Economy*. New York: UN Department of Economic and Social Affairs.

UN ESCAP. 2000. *State of the Environment in Asia and the Pacific*. Manila: Asian Development Bank.

Weder, B. 1999. *Model, Myth, or Miracle? Reassessing the Role of Governments in the East Asian Experience*. Tokyo: United Nations University Press.

World Bank. 2000. *Entering the 21st Century – World Development Report 1999/2000*. New York: Oxford University Press.

World Bank. 2001. *World Bank Atlas 2001*. Washington, DC: World Bank.

Case studies of pesticide management

2

Malaysian perspectives on the management of pesticides

Abdul Rani Abdullah and Saraswathy Sinnakkannu

Malaysian rice production meets about 80 per cent of the country's rice consumption needs, where paddy production has been estimated at 1.3 million tonnes per year. The consistently high yields of paddy have been attributed in part to the widespread application of chemical pesticides, including herbicides, insecticides, and fungicides. Malaysia still relies largely on less-expensive, long-established pesticides, as do the majority of other Asian nations. Incidentally, rice has also been identified as being particularly susceptible to the adverse impacts of the often indiscriminate and intensive use of chemicals and pesticides. Further, pesticide residues have been detected in water, soil, and fish in the paddy-field ecosystem as well as the general environment, indicative of potential detrimental impacts to many non-target organisms, including the human population.

In general, concerns about changes in the status of rice pests and diseases and negative impacts on the environment and human society have increased. Current estimates put global losses due to animal pests, diseases, and weeds at US$300 billion annually. This is equal to around 30 per cent of potential global food, fibre, and feed production (UN ESCAP 1994). Yields of intensive rice production have been levelling off and the resources base for rice production has shown signs of accumulated stresses. Pests not only cause quantitative losses but also affect the quality of agricultural produce, which may become noticeable through a deterioration of taste, flavour, fibre strength, processing characteristic, seed germination, etc. With growing demand for rice, which is a staple food, and with limited opportunities to open up new areas for agriculture, it has become necessary to adopt intensive crop management.

As an innovative approach to circumvent the problems associated with regulating the production and applications of pesticides, integrated pest management (IPM) strategies involving a variety of biological, physical, and chemical methods are promoted. These IPM technologies have been implemented in Malaysia with varying degrees of success. The Malaysian government has launched a multimedia strategic extension campaign targeted at the adoption of a holistic approach to overcome grassy weed problems in direct-seeded rice. As a result of this campaign, a remarkable decline in herbicide usage has been observed. It is interesting to note that several NGOs in Malaysia are also playing a complementary role publicizing the IPM strategies. Particularly, their contribution in the form of basic consumer education as well as organizing forums and seminars on the misuse of pesticides and advocating safe and sustainable pest control methods is somewhat novel in the East Asian context.

At the same time, managing the crop protection chemical industry in Malaysia – which comprises about 140 multinational and local companies – is indeed a daunting challenge. Because the impacts of pesticides on human health arise from occupational exposure and ingestion of tainted foodstuffs, legislation to safeguard human health and the environment from pesticide residues must also be put in place. In this regard, a variety of government agencies such as the Department of Agriculture, Department of Environment, and Department of Health have been given the task to administer and enforce regulatory policies and laws pertaining to all aspects of pesticide use. Examples of such legislation include the Pesticide Act, the Environmental Quality Act, and the Food Act/Regulations.

Significance of rice production in Malaysia

Rice is the major source of food for as much as 60 per cent of the world population (Mabett 1991). World trade in rice was projected to expand by 4 per cent a year to 18.9 million tonnes by 2000, substantially faster than the rate in the 1980s (*Malaysia Agriculture Directory and Index* 1998). Exports by developing countries are expected to increase by more than 50 per cent, while exports by developed countries are expected to decline due to the anticipated reduction in export subsidies. World rice stocks are expected to rise 2.5 per cent a year to reach 68 million tonnes, or 17 per cent of the projected annual consumption (*ibid.*).

Rice is the predominant food crop in the tropics (Claro 1998). Malaysia grows about 80 per cent of her rice consumption needs, occupying about 467,300 hectares of land in the 1993–1994 main season, of which 362,628

Table 2.1 Statistics of rice production in Malaysia's main granary areas, 1996

Granary area	Area planted (ha)	Area harvested (ha)	Average yield (kg/ha)	Paddy production (tonnes)	Rice production (tonnes)
MADA	193,223	193,115	4,328	836,359	543,633
KADA	53,892	53,601	3,676	198,117	128,776
Kerian Sg Manik	45,099	45,099	3,106	140,086	91,056
Barat Laut Selangor	35,878	35,878	4,314	154,778	100,606
P.P.P.B. P. Pinang	19,438	19,437	4,037	78,473	51,007
Seberang Perak	17,128	17,127	3,925	67,232	43,701
Besut (Ketara)	9,063	8,832	4,783	43,352	28,179
Kemasin Semerak	8,732	8,686	3,028	26,438	17,184
Total/mean	382,453	381,775	4,039	1,544,834	1,004,142

Source: DOA (1996)

hectares were wet paddy and 104,649 were dry paddy. Figures for area under rice cultivation and the overall production during 1996 are presented in Table 2.1. This area had increased to 535,057 hectares by the year 2000 (DOA 2000), as shown in Table 2.2. It should be pointed out that the cultivated area varies from year to year and is sometimes under conversion to other crops, particularly oil palm and for commercial and residential uses.

In 1996 the *Paddy Statistics of Malaysia* prepared by the Department of Agriculture (DOA) provided estimates of paddy production in the

Table 2.2 Rice hectareage in Malaysia, 2000

State	Area planted (ha)
Johor	2,591
Kedah	52,942
Kelantan	75,031
Melaka	2,004
Negeri Sembilan	1,702
Pahang	6,801
Perak	82,752
Perlis	51,415
Pulau Pinang	28,591
Selangor	37,888
Terengganu	22,370
Sabah	40,089
Sarawak	130,881
Total	535,057

Source: DOA (2000)

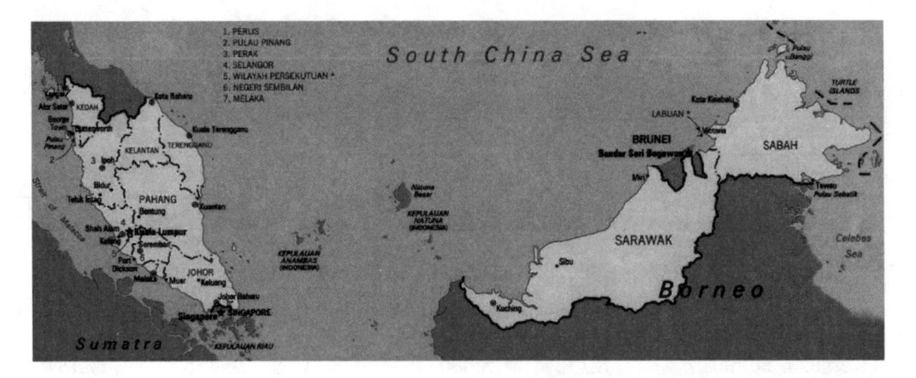

Figure 2.1 Map showing states within Malaysia

country, as shown in Table 2.1; a map showing the states within Malaysia is provided in Figure 2.1. In general, peninsular Malaysia accounts for 80 per cent of the national paddy production, Sabah 9 per cent, and Sarawak 10 per cent. Paddy production will continue to be concentrated in the eight granary areas, where farmers are encouraged to produce high-quality rice to benefit from higher market prices. Outside the granary areas, farmers are encouraged to switch to more profitable crops. The main rice-growing areas are the Muda Agricultural Development Authority (MADA) scheme located in Kedah/Perlis and the Kemubu Agricultural Development Authority (KADA) in Kelantan, which accounted for 50 per cent and 14 per cent of the total planted area in 1996, respectively.

Relationship between pesticide industry and rice production

Globally, more pesticides are used on rice than on any other single crop; an estimated 12 per cent of the world's US$30.3 billion crop protection market is spent on the chemical control of weeds, pests, and diseases in rice, followed by maize (11 per cent), and cotton (9 per cent) (Warrell 1997). Losses of the rice crop to insects have been estimated to be six times higher than similar losses in wheat, rye, oats, and barley of the temperate zone (Bunton 1991). A total of approximately 46 per cent of the potential global rice crop is estimated to be lost through insects (28 per cent), diseases (8–10 per cent), and weeds (10 per cent) (*ibid.*).

According to the Malaysian Agricultural Chemical Association (MACA), the crop protection chemical industry in Malaysia comprises about 140 companies, consisting of both multinational and local firms; most of them are involved in formulation and/or trading of pesticides (MACA 1997). The majority of the pesticides used in Malaysia are ap-

Table 2.3 Estimate of the pesticide market in Malaysia (US$ million)[a]

Pesticide class	1990	1991	1992	1993	1994	1995	1996	1997
Herbicides	104.5	92.0	84.0	80.0	80.4	88.0	90.8	97.8
Insecticides	17.1	16.0	16.4	15.6	16.4	17.2	18.8	20.9
Fungicides	5.8	5.2	5.2	5.2	5.6	6.0	6.4	6.5
Rodenticides	4.2	4.0	4.8	4.0	4.4	4.4	4.4	5.2
Total	131.6	117.2	110.4	104.8	106.8	115.6	120.4	130.4

Source: MACA (1996 and 1997)
a. End-user value at a constant exchange rate of US$1 = RM2.5

plied in the rubber, oil palm, and rice sectors of agriculture. As can be seen in Table 2.3, herbicides account for the majority of the total pesticide market. In 1997, the total market value based on end-user values was estimated to be RM326 million (around US$85 million). Herbicides constituted 75 per cent, followed by insecticides (16 per cent), fungicides (5 per cent), and rodenticides, molluscicides, and others (4 per cent) (Ho 1998). In Malaysia, the import of pesticides increased from 8,589 tonnes in 1994 to 9,405 tonnes in 1995 (Statistics Department of Malaysia 1995). On average, a Malaysian farmer spends around RM400 on pesticides for an acre annually.

The pesticide industry is one of the most important supporting industries for agriculture in Malaysia. The economic benefits of pesticide use in producing high crop yield, as well as the role of pesticides in the control of disease-borne pests, are undeniable. As an illustration, a weed survey in the Muda area in Kedah indicated that the number of farmers who reported the incidence of *Echinochloa complex* infestation declined from 77 per cent in 1987 to 25 per cent in 1993, while the number of farmers who reported *Leptochloa chinesis* infestation reduced from 46 per cent to 32 per cent over the same period. The effectiveness of weed control reflected an improvement in rice yield performance. In the first season of 1987, the gross yield was only 2.7 tonnes/ha, whereas by the first season of 1993 the gross yield had escalated to 4.3 tonnes/ha, indicating an increase of 59.3 per cent over a period of six years (Ho *et al.* 1996).

While newer pesticide formulations are being gradually adopted, Malaysia still relies largely on less expensive, long-established chemicals, as do other rice-producing nations such as China and India and most other South-East Asian nations. Hence, carbamates such as BPMC, carbaryl, and carbofuran have been the mainstay for many years for the control of the brown planthopper, *Nilaparvata lugen*, while organophosphates

(OPs) such as diazinon, fenitrothion, and malathion (sometimes in combinations with carbamates) have been used to combat stem borers such as *Chilo supressalis* (Warrell 1997). Table 2.4 lists examples of commonly used pesticides applied in Malaysian paddy fields.

The warm, tropical conditions in Malaysia, characterized by high rainfall (2,100–2,500 mm/year), humidity (52–96 per cent), and warm temperature (21–35°C), are conducive to crop and weed growth (Chee *et al.* 1991). Yield depression and crop losses due to strong competition for nutrients, moisture, light, and allelopathy from weeds can be very significant. Weeds also affect the perennial crops indirectly by their thick coverage, which makes accessibility for manuring, pruning, and crop protection difficult. Severe weed infestation also causes irregular and poor harvest (Ho 1997). Hence, it is not surprising that herbicides have consistently accounted for more than 70 per cent of the total pesticide market in Malaysia.

Marketing of pesticides in Malaysia

Any pesticides manufactured or imported into the country must be registered with the Pesticide Board, which was established to implement the Pesticides Act 1974 (Section 6.2). There are at least 146 companies registered with the Pesticide Board to manufacture, distribute, or sell the chemicals in the country (Pesticide Board Malaysia 1996). Pesticides like 2,4-D, MSMA, sodium chlorate, dalapon, glyphosphate, and paraquat are either wholly or partially manufactured locally. Others are imported as technical material and locally formulated for use. The distribution of pesticides is largely handled by the commercial sector. The distribution may be categorized as follows:

- short distribution, which goes from supplier direct to the retailers
- long-chain distribution, whereby the supplier sells to distributors who sell to various retailers
- from suppliers to stockists, who in turn distribute to the dealers, government agencies, and estate agencies, who distribute to the end users.

The government under the Farmers' Organization Authority also has retail shops throughout the country, whose main function is to distribute pesticides. Pesticides may also be distributed free of charge as a subsidy to specific government projects for the farmers (Ismail 1998). Most farmers depend on their neighbourhood dealers, shopkeeper, friends, or government officers for advice, as the majority of these farmers have little knowledge about the types and safe use of pesticides.

More often farmers are not briefed about the potential dangers to health and long-term drawbacks of pesticides. This is partly due to the retailers themselves being ignorant of such problems, as well as their lack

Table 2.4 Commonly used pesticides in paddy fields

Type of pesticide	Chemical name	Trade name	Type of pest
Herbicide	2,4-D Butyl ester	Rumputox (45%), Solartox (45%), Tox-special (45%)	*Eichhornia crassipes*, *Monochoris* spp, *Jussiaea* spp, *Sagittaria guyanensis*
	2,4-D Dimethylamine	2,4-D Amine 600 (60%) Acmine (48%), Amine CP500 (50%), Amine Super (69.5%)	*Fimbristylis* spp, *Dicranopteris linearis*, *Borreria latifolia*, *Amaranthus* spp, *Mimosa pudica*
	Glyphosphate Isopropylamine	Broadcut (13.6%), Ace-round (29%), Achieve (41%), Imej 62 (62%)	*Imperate cylindera*, *Ottochloa nodosa*, *Cyrtococcum oxphyllum*, *Mimosa invisa*
	Propanil	Kensolo (26.8%), ACM Propanil (34%), Wham EZ (42%), Lintar (60%)	*Ischaemum rugosum*, *barnyard grass*, *Panicum repens*, *Scirpus grossus*
	Paraquat Dichloride	Sekuat (13%), Terquat (19%) Telezone (25%), Uniquat (42%)	*Paspalum conjugatum*, *Scieria* spp, *Eleusine indica*, *Cloemma hirta*
Insecticide	Acephate	Ace 750 (75%), Orthene 75S (75%), Impact 75 (75%), Compete 75 (75%)	*Scotinophora coartata*, *Cnaphalocrosis*, *Nilaparvata lugens*, *Leptocorisa* spp
	BPMC	Bulldock 25 EC (25%), Bena 555 (50%), Bassa 50 EC (50%), Vitagro 50EC (50%)	*Valanga nigriconis*, *Nephotettix* spp, *Sogatella furcifera*, *Scotinophora coartata*
	Carbaryl	Carbacide (5%), Avin 85 (85%), Boly 85 (85%), Rebel 85 (85%)	*Chilo polychrysa*, *Tryporyza incertulas*, *Spodoptera litura*, *Sesamia inferens*
	Endosulfan	Endotox (32.9), Kendan (33%), Premofan 35EC (35%), Endocel Tech (94%)	*Cnaphalocrosis* spp, *Tryporyza incertulas*, *Spodoptera litura*, *Sesamia inferens*
Rodenticide	Brodifacoum	Matikus, Talon, Rm Maiz	*Rattus argentiventer*, *Bandicota indica*
	Warfarin	Hapus Tikus, King Kong Rat Killer, Tikumin, Rid Rat	*Rattus argentiventer*, *Bandicota indica*
	Zinc phosphide	Micekil, Ratcon, Rid-all	*Rattus argentiventer*, *Bandicota indica*

Source: Pesticide Board Malaysia (1996)

of adequate knowledge of pesticide usage. Consequently, farmers generally use more pesticides than they should, under the misconception that more chemical pesticides used would mean more pests eradicated. They deduce that their crop yield would be higher with the use of more pesticides, leading to an anticipated increase in their incomes.

Impacts of pesticide use on environment and society

The primary concern for the presence of pesticide residues in the environment arises from their toxicity to all living organisms. Hence, in addition to the intended target pests, non-target organisms including humans are exposed to the toxic effects of pesticides, which may include mutagenic, carcinogenic, and teratogenic effects. Other effects – particularly bioaccumulation in the food chain – are mainly associated with the organochlorine (OC) class of compounds. The adverse effects of elevated pesticide residues in water, soil, and crops to man, domestic animals, wildlife, and the environment in general are well recognized and documented.

In tropical countries such as Malaysia, the rice crop has been identified (together with vegetables) to be particularly susceptible to the negative impacts of pesticides (ADB 1987). This is attributed to the often indiscriminate and intensive use of pesticides associated with these crops. The problem is exacerbated by the inadvertent destruction of natural predators, as well as the emergence of resistant pest strains, the consequence of which is again application of increasingly larger amounts of the chemicals.

In recent years, ultra-low-volume techniques employing mist for herbicide application have been used for general weed control in Malaysia. This technique was popular as it was capable of overcoming problems related to labour shortage and the increasing cost of weed control. Malaysian farmers usually spray pesticides using a backpack of equipment, clad only in T-shirts, shorts, and slippers that offer little protection. When powered sprays and fogging machines are used, the only protection against inhalation of toxic fumes is usually just a handkerchief or towel to cover the mouth. The application of pesticide with backpack sprayers has been proven to be inefficient, as only about 20 per cent of the spray reaches the plants and less than 1 per cent of the chemical contributes to pest control, resulting in wastage and contamination of the environment (*Malaysian Agricultural Directory and Index* 1996).

Impact on non-target organisms

The use of pesticide for specific pest control often leads to unintended effects on non-target organisms. In addition to the destruction of natural enemies, pesticides are known to affect a number of other organisms, such as other invertebrates, birds, and mammals (although most of the currently used pesticides have relatively low toxicity to mammals, a primary consideration in the registration approval process). In addition to the toxic effects of pesticides to paddy-field fish, the uptake and subsequent accumulation of pesticide residues in fish tissues result in the transfer of the chemical through the food chain.

Aquaculture production of fish in paddy fields is a common practice. Common species belong to the *Anabantidae*, *Channidae*, *Siluridae*, and *Cichlidae* families. During flooding, fish normally migrate into paddy fields, and when the rice matures the fish are harvested. This aquatic environment is generally considered to be the main repository for pesticides discharged to the environment. The existence of fish in the paddy fields has a beneficial effect on paddy yields. Some species feed on algae and recycle nutrients into the environment via their faeces. Pest population is also reduced because some insects' larvae are eaten by certain varieties of fish. In West Malaysia some 362,000 hectares are under wet paddy cultivation (which hold water for about seven months a year) and are most suitable for fish production. Fish incomes account for about 25–50 per cent of the paddy farmer's income (Ibrahim 1992). Fish is also an important and cheap source of protein, contributing 70 per cent of the total protein intake of the Malaysian diet (Umakanthan 1983; Ismail 1998).

Table 2.5 provides an indication of the extent of contamination by organochlorine pesticide residues in water, soil, and fish of the paddy-field ecosystem in Malaysia. Pesticides which have been banned or whose use has been restricted, such as dieldrin, endrin, and DDT, continue to be detected due to their persistent character. In a study by Tan and Vijayaletchumy (1994a), hexachlorocyclohexane (HCH), DDT, heptachlor, and dieldrin were found in the water of many of the rivers flowing through rice-growing areas. Endosulfan, an insecticide with a well-documented piscicidal activity, was also commonly detected. HCH, heptachlor, aldrin, and endosulfan were also detected in sediments taken from a rice-growing area in Sungai Bernam and Sungai Selangor (Tan and Vijayaletchumy 1994b). Significant levels of both HCH and endosulfan are due to the current usage of these chemicals in the paddy fields. In a survey conducted by Mustafa (1997) on polished rice taken from various locations in peninsular Malaysia, OP pesticides such as diazinon, chlorpyrifos, and malathion, and the OC pesticide heptachlor were not detected in the samples with the exception of the samples taken in the state of Kedah, where a concentration of 7.59 ng/g fenitrothion was found in addition to the detection of o,p-DDE, p,p-DDE, and p,p-DDT.

Use of excessive amounts of pesticides in Malaysia has led to a number of cases where fish kills have been attributed either directly or indirectly

Table 2.5 Pesticide residues in water, soil, and fish in paddy fields in Malaysia

Location	Survey year	Chemical	Concentration[a] (water ug/L) (sediment/biota:ng/g)	References
Krian river basin	1981	Dieldrin	ND–0.5	
(Water)		Beta-HCH	0.1–0.9	Meier, Fook, and Lagner (1983)
		Gamma-HCH	ND–0.6	
		Aldrin	0.1–1.8	
		p,p′-DDT	ND–1.6	
(Soil)		Dieldrin	ND–4.7	
		Beta-HCH	0.6–0.8	
		Gamma-HCH	0.4–0.8	
		Aldrin	0.1	
		p,p′-DDT	1.0–4.0	
(Paddy-field fish)		Dieldrin	6.6–24.9	
		Alpha-chlordane	2.8–17.1	
		Beta-HCH	3.3–8.2	
		Aldrin	0.3–8.2	
		p,p′-DDT	0.3–1.1	
			2.2–6.0	
Tanjung Karang	1982	Alpha-HCH	0.5	Soon and Hock (1987)
(Water)		Gamma-HCH	0.1	
(Gouramy)		Alpha-HCH	18–58	
Trichogaster trichopterus		Gamma-HCH	10–100	
		Alpha-endosulfan	5,130	
		Beta-endosulfan	1,700	

Perai river basin, Kedah (Water)	1990–1991	t-HCH	0.065	Tan and Vijayaletchumy (1994a)
		t-endosulfan	0.110	
		Heptachlor	0.026	
		t-DDT	0.048	
		Dieldrin	0.015	
Bernam river basin, Perak (Water)		t-HCH	0.320	
		t-endosulfan	0.062	
		Heptachlor	0.120	
		t-DDT	0.190	
		Dieldrin	0.039	
Kelantan river basin, Kelantan (Water)		t-HCH	0.0005	
		t-endosulfan	ND	
		Heptachlor	0.001	
		t-DDT	0.001	
		Dieldrin	ND	
Bernam river basin, Perak (Sediment)	1992–1993	HCH	3.520	Tan and Vijayaletchumy (1994b)
		Heptachlor	1.275	
		Endosulfan	0.960	
		Aldrin	0.045	
Two unspecified locations in Kedah (Polished rice)	1997	p,p-DDT	4.70, 15.89	Mustafa (1997)
		O,P-DDE	5.90, 16.79	
		p,p-DDE	ND, 8.03	

Source: Pesticide Board Malaysia (1996)

ND = not detected; a = average/range, t-HCH = alpha, beta, and gamma HCH; t-endo = endosulfan I + endosulfan II; t-DDT = p,p′-DDE + p,p′DDT to pesticide residues. Frequent reports of fish kills during the 1960s and 1970s initially prompted research into the toxic effects of pesticides using indigenous species of fish (Yunus and Lim 1971; Moulton 1973; Soon and Hock 1987). Toxic effects of pesticides on paddy-field fish have also been observed based on physiological and biochemical changes as well as behavioural alterations. Hence, the anticholinesterase effects of OP insecticides were noted in paddy-field fish such as carp fingerlings exposed to sublethal doses of methamidophos (Yew and Sudderuddin 1979) and red tilapia (hybrid *Tilapia mossambica x T. nilotica*) exposed to a lethal dose of malathion (Sulaiman, Abdullah, and Ahmad 1989). In a field study, bagrid catfish (*Mystus julio*) in an irrigation canal of a paddy field were found to exhibit depressed acetylcholinesterase activity attributed to the use of OP insecticides (Abdullah, Sulaimau, and Mohamed 1994).

Besides fish, other biota have also been affected by pesticide residues in the paddy-field ecosystem. Of particular concern are toxic effects to natural predators of pests. Frogs and snakes are commonly observed to be affected by pesticide application in the paddy field (Soon and Hock 1987). Local bird experts have also reported that in areas of large-scale paddy cultivation the number of migratory egrets and herons has declined sharply over a decade or so, coinciding with a marked increase in the application of pesticides. Another small but no less important assemblage of species that might be at risk due to the exposure to pesticides are migratory birds of prey (the greater spotted eagle and the black kite) which frequent paddy-field habitats and fresh-water waders like snipes and sandpipers (CAP 1985).

Development of resistance

With the continued use of pesticides for pest control, pest species often develop resistance to the pesticide in use. The problem is best illustrated in numerous cases of resistance demonstrated in insect pest species. Resistance to insecticides in the green leafhopper population has been recognized since the early 1960s, as exemplified by resistance to OPs such as malathion and methyparathion. The subsequent use of carbamates such as carbaryl, BPMC, carbofuran, and isoprocarb also saw the emergence of resistant strains in the 1970s (Mabbett 1997). Insect resistance to insecticides can arise by three mechanisms that include biochemical, physiological, and behavioural processes (Mohan 1987).

Kim (1996) reported that a shift to resistant weeds could evolve either slowly or rapidly, depending on both cultural and ecological conditions. The key factors that favour the rapid development of a herbicide-resistant weed population are as follows:

- use of a highly effective herbicide which controls most of the susceptible biotypes (selection pressure), causing a rapid population shift in favour of the resistant biotypes
- the number of naturally occurring resistant biotypes within the native weed population, which increases under selection pressure
- monoculture cropping with repeated usage of the same herbicide on the same crop; this creates an ideal environment for herbicide resistance to spread.

Controlling weeds with continuous 2,4-D application has led to the emergence of herbicide-resistant biotypes in the paddy fields. In the Muda area, a 2,4-D resistant biotype of *Fimbristylis miliacea* from the family *Cyperaceae* was first detected in 1989, where 2,4-D has been seasonally applied since 1975 (Ho 1992). The development of resistance has prompted farmers to use extremely toxic and non-registered pesticides which have been smuggled into the country, or to mix pesticides to boost their effectiveness, resulting in a more serious pesticide problem in the country.

Impact on human health

While the major route of contamination in the general population is via the ingestion of food tainted with pesticide residues, the more direct mode of exposure involves occupational and accidental exposure which can occur in the manufacture of the chemicals or during their application.

A survey by Sahabat Alam Malaysia (Friends of the Earth Malaysia) in 1984 revealed some alarming facts (SAM 1984).

- Workers are not provided with any protective clothing so they use their own regular clothing. These clothes, contaminated with pesticides, are washed together with other clothing.
- Workers do not wash themselves thoroughly after handling the pesticides because they are not provided with adequate facilities.
- Workers do not have proper equipment to mix the chemicals, so they use their bare hands.
- The estate management does not provide any training or information with regard to the potential dangers of pesticide use to the labourers.
- In the absence of any laws regulating the safe and proper disposal of chemical containers, the workers use these containers for their daily use – for storing water, cooking oil, and other consumption items.
- The workers' ignorance of the potential danger of pesticide use can also be illustrated in the storage of dangerous pesticides, which are kept close to food.

Improper handling, lack of protective measures, and lack of knowledge among the farmers have led to widespread cases of poisoning across the country. In 1991, 893 cases of poisoning were reported in government hospitals throughout Malaysia. Of these, 278 (31.1 per cent) were fatal (Ministry of Health 1991). Out of 916 cases in 1994, 156 (17.03 per cent) were fatal (Ministry of Health 1994). Forty per cent of human poisoning cases are due to pesticides, of which paraquat accounted for about 70 per cent of all the cases reported over the period 1979–1988. Although pesticide poisoning is often associated with suicides, accidental exposures at work accounted for about 11 per cent of the 6,554 poisoning cases reported between 1989 and 1994 (Mokhtar 1996). In addition to farmers, workers in pesticide manufacturing plants, distributors, and retailers involved in repackaging pesticides, workers who dilute and mix the concentrates, those who load and maintain spraying machinery, and even consumers who consume the farm produce are all constantly exposed to the dangers of pesticides. The highest number of poisoning incidences was due to herbicide application alone (24.8 per cent), followed by insecticides (14.7 per cent) (Kamaruddin 1998). Farmers rarely experienced hazardous symptoms due to the use of rodenticides or fungicides.

The appropriate legislation governing the safe handling of pesticides deemed to be highly toxic is encompassed in Section 57(1) of the Pesticides Act 1974. This Act provides for the measures to be taken and the practices to be followed or avoided by manufacturers and employers for the protection, safety, and well-being of their workers engaged in the manufacture or handling of pesticides, and was enforced under the Pesticides (Highly Toxic Pesticides) Regulations 1996. The regulations came into force on 1 April 1996, and their main objective is to place greater accountability on the employers and workers using or handling certain highly toxic pesticides so as to minimize the risks posed by the chemicals.

A study to assess the knowledge and practices related to pesticide handling and use among paddy farmers in Muda area in 1993 revealed that 71.5 per cent of the farmers are over 39 years old, 61.7 per cent have only primary education, and 66.6 per cent of them practise individual farming (Jamal *et al.* 1994). Queries regarding farmers' knowledge revealed that 62.4 per cent were aware that there is a need for a special storage place for pesticides, 58.6 per cent were aware of the possibility of poisoning occurring during pesticide mixing, 55.5 per cent practised proper disposal methods for pesticide containers, and 70.5 per cent of them had experienced early symptoms of pesticide poisoning such as headaches, skin diseases, and nausea. Most farmers (70.5 per cent) were aware that the main route of entry of pesticide into the human body is through inhalation, but few (19.4 per cent) were aware that pesticides could enter the body through the skin. Improper farmers' practices such as the use of increased dosages (61.2 per cent) and the mixing of pesticides (57.8 per cent) were also noted to be quite widespread. Both farmers' knowledge and their practices with respect to personal protective equipment and personal cleanliness were found to be very poor.

In recent years there has been increasing concern over environmental pollutants that can mimic oestrogenic effects, often referred to as endocrine disruptors. Adverse effects from exposure to these chemicals arise due to interference of the chemical with natural hormones in the bodies of the affected organisms; the hormonal system is generally responsible for the maintenance of homeostasis and the regulation of developmental processes (Kavlock *et al.* 1996). Reported adverse effects in wildlife include population declines, increases in cancers, reduced reproductive function, and disrupted development of immune and nervous systems (Jimenez 1997). In addition to environmental pollutants such as PCBs, bisphenols, and dioxins, numerous studies have indicated that some pesticides also elicit toxic responses which are related to disruptions of the endocrine system (Rivas *et al.* 1997). These include DDT and its metabolites, aldicarb, atrazine, benomyl, endosulfan, lindane, malathion, mancozeb, maneb, and permethrin amongst others (Gascon *et al.* 1997). Many of these chemicals are widely used in Malaysia, including in the paddy fields.

While there is clear evidence to suggest that some environmental pollutants such as PCBs, bisphenols, and pesticides can alter the endocrine function in fish, birds, and mammals, including humans, cause-effect relationships have not always been clearly established. Hence, studies relating breast cancer to exposure to DDT as an endocrine-disrupting chemical have been controversial (Rivas *et al.* 1997). Certainly, studies designed to establish a relationship between exposure to endocrine-disrupting chemicals and harmful effects on development and hormonal homeostasis in various animal species under controlled conditions have been more numerous than those involving humans. At present studies directed towards human exposure include the establishment of cause-effect relationships, epidemiological analysis of health effects on exposed population, and the establishment of possible mechanisms of action of the chemicals (*ibid.*).

Malaysian approach to pesticide regulation – Policies and laws

Because of the toxic nature of pesticides and the potential hazards they present to man and the environment, laws and regulations are in place to regulate their use. Government agencies such as the DOA and the Department of Environment play important roles in administering and enforcing these regulations. The relevant policies and legislation pertaining to pesticides in Malaysia are the National Agricultural Policy, the Pesticides Act, the Environmental Quality Act, and the Food Act/Food Regulations.

National Agricultural Policy 1992–2010

The National Agricultural Policy was launched in January 1984 to provide the framework for a balanced and sustained rate of growth in the agricultural sector. It sets out the guidelines for agricultural development over the medium term, highlighting the importance of the sector to the economy, the constraints encountered, and the broad strategies to be adopted (Abdul 1993). The National Agricultural Policy (NAP) 1992–2010 emphasizes efforts to increase farmers' income by raising productivity and changing crops as well as agricultural management practices. It includes plans to group farmers into large-scale mini-estate operations to attain economies of scale in labour, management, and production. Small-scale farmers and smallholders are to be consolidated into larger holdings. All these proposals assume that farmers will opt for cash crops and high-value activities to maximize farm incomes (*ibid.*).

The dampened performance of agriculture against the backdrop of increasing competition, escalating protectionism, and the relative ineffectiveness of agricultural trade arrangements has prompted the government to review the NAP critically and to offer an alternative plan for the sustained development of the agricultural sector. The NAP 1992–2010, which was unveiled by the Prime Minister in 1993, attempts to consolidate the earlier policy in the context of accelerating the transformation of agriculture into a modern, highly commercialized, and stable sector.

Pesticides Act 1974

The Pesticides Act 1974 is a comprehensive Act aimed at controlling the various aspects of pesticide use in the country. To implement the Act, the Pesticide Board comprising members from related government agencies was established. The DOA was entrusted with the task of providing the secretariat to the board in the implementation of the Act. The main objective of the Pesticides Act 1974 is to ensure that pesticides imported, manufactured, and sold in the country are effective for their intended use and have no unacceptable adverse effects on crops, humans, and the environment. For effective implementation, six main sets of rules and regulations have been formulated and enacted.

- Pesticides (Registration) Rules 1976 – a pesticide is evaluated among other aspects on the product chemistry/quality, toxicology, efficacy/

use, residue chemistry/effects, environmental fate, packaging/labelling, and risk/benefits.
- Pesticides (Labelling) Regulation 1984.
- Pesticides (Importing for Educational and Research Purposes) Rules 1981.
- Pesticides (Licensing for Sale and Store for Sale) Rules 1988.
- Pesticides (Highly Toxic Pesticides) Regulation 1996.
- Pesticides (Advertisement) Regulation 1996.

To register a chemical an appropriate form has to be completed, providing relevant information such as the physico-chemical properties of the pesticide, toxicity data, and efficacy and residue levels. A sample of the pesticide is also required for analysis. A pesticide that does not gain registration by the Pesticide Board cannot be sold in Malaysia. This means automatically that any pesticide which is not registered is banned in Malaysia. The registration is valid for five years, after which the registered pesticide is reassessed for its continued use. In this way several pesticides have either been deregistered or their use restricted following reassessment – a procedure which takes into consideration safer alternatives and reported abuses among other factors.

The use of pesticides deemed to be highly toxic is further governed by the Pesticides (Highly Toxic Pesticides) Regulation 1996, particularly with respect to specific handling restrictions, ensuring that the workers handling these chemicals do so with the utmost care. Employers are required to provide adequate training to the workers, who have to be medically fit and are only permitted to work for a maximum of eight hours a day. A strict record is required to be maintained, including details such as the number of hours worked, the amounts of pesticide used, and the method of application. The workers are also required to wear proper protective clothing and annual medical examinations are to be carried out on them. Furthermore, these regulations dictate that in the case of female workers, only those who are not pregnant or lactating are allowed to handle the pesticides. Other requirements include the safe and proper storage of the chemicals and the safe disposal of empty containers. The handling of pesticides such as paraquat, monocrotophos, and calcium cyanide is subjected to these regulations.

Environmental Quality Act 1974 (amended 1985)

The main regulatory instrument that controls the handling and disposal of pesticide wastes in Malaysia is the Environmental Quality (Scheduled Waste) Regulation 1989, which was formulated under the Environmental Quality Act 1974 (amended 1985). This Act comes under the Department of Environment in the Ministry of Science, Technology, and the Environment.

Pesticide wastes are subjected to the requirements of the regulation, covering:

- a notification system for waste generators
- a licensing system for treatment and disposal facilities
- a waste inventory
- labelling requirements
- responsibilities of waste generators
- treatment facilities
- recycling
- transportation and final disposal site operators.

The estimated quantity of toxic and hazardous wastes generated in 1987 by the pesticide industry in Malaysia was 251 cubic metres. Most of this waste came from pesticide formulation plants. In addition, about 70,000 pesticide-contaminated containers of various sizes and forms are produced annually. A non-quantifiable amount of pesticide wastes is also produced by pesticide end users (Ibrahim 1992).

Food Act 1983/Food Regulations 1985

The Food Act 1983/Food Regulations 1985, enforced under the Ministry of Health, control pesticide residues in food. Maximum residue limits for pesticide residues are stipulated in this Act. The use of pesticides can result in serious and widespread health hazards and threats to the environment. It is therefore vital that the level of pesticide residue in food is closely monitored and the necessary control measures implemented. The Act also provides for punitive action against those who misuse pesticides and by their action lead to unacceptable residues in food.

In reality, contamination of food due to pesticide residues continues to occur although advice, warnings, and court prosecutions are undertaken. Over 6,000 samples of various foods are analysed annually by the quality control unit at the Ministry of Health, of which up to 7.7 per cent have been found to contain pesticide residues exceeding the minimum requirement levels (Ministry of Health 1994). Testing of rice for local consumption for pesticide residues is also carried out by the quality control unit. Testing of food for pesticide residues is conducted in accordance with the International AOAC Official Methods of Analysis 1984 (Chin 1998).

Role of Malaysian government

A multimedia strategic extension campaign was launched in 1989 by the DOA to inform, motivate, and persuade the target audience on the adoption of a holistic approach to overcome grassy weed problems in direct-seeded rice. The principal objective of this campaign was to create awareness among the rice farmers regarding the beneficial role of biological agents, and the importance of conserving the predators, egg parasitoids, and pathogens for sustainable rice production. Continuous implementation of the strategic extension campaign in the period 1990–1994 showed remarkable results. The dry-seeded first-season rice yields increased steadily to 4.6 tonnes/ha in 1995 in the Muda area in Kedah. It is noted that over the same period the usage of herbicides declined. This is attributed to the fact that the standard of land preparation improved significantly (Ho and Zainuddin 1995).

A variety of weed control options have also been developed from multi-locational farmer-managed trials. These include avoidance of an excessively high seeding rate to minimize pest and disease infestation, use of weed-free certified seeds, carrying out regular hand weeding, and reporting the emergence of new weed species and incidences of herbicide resistance immediately to the extension agents.

In addition to the strategic extension campaign, the DOA together with relevant agricultural authorities have focused on a range of activities for the improvement of farmers' pest management.

- Strengthening group action at the field level through rice mini-estate management and adopting a farmer-to-farmer extension approach.
- Expanding the services of crop protection clinics at farm localities.
- Encouraging farmer-based surveillance to complement and supplement the current pest surveillance and forecasting system implemented by the Department of Agriculture.
- Training the workers on proper pesticide application techniques.
- Conducting farmer participatory research on weed management.
- Studying the non-rice habitats in the agro-ecosystem and assessing their importance as a refuge for beneficial organisms.
- Evaluating the long-term effects of rice-fish culture and rice-duck culture as alternatives for sustainable pest management.
- Monitoring of pesticide residues in the rice ecosystem.

Integrated pest management as an innovative solution

The philosophy and principles of integrated pest management (IPM) have been known for nearly three decades. As a means to promote more judicious use of pesticides in the paddy fields, IPM strategies involving a variety of biological, physical, and chemical methods have been adopted with varying degrees of success.

Malaysia was among the seven countries which participated in the FAO/UNDP programme for integrated pest control in rice initiated in

1975 (Majid, Lim, and Booty 1984). The primary objective was to improve the overall pest control and rice production capability by the introduction and implementation of specific strategies in research and application of IPM, emphasizing among other aspects training and field demonstrations, pest forecasting and surveillance, pesticide application, biological control, stable pest resistance, and environmental impacts (*ibid.*).

In peninsular Malaysia the project was implemented in all the major rice-growing areas, commencing in June 1980. The Crop Protection Unit of the DOA was responsible for implementing and coordinating the programme, while the Malaysian Agricultural Research and Development Institute (MARDI) provided the necessary research activities. The FAO provided technical support and training. Encouraging progress was achieved, particularly in pest surveillance and training. A positive impact was also achieved in awareness among farmers of the concept and acceptance of IPM. Hence farmers were made to be more aware of the role of natural enemies, leading to more judicious use of pesticides. A commonly accepted form of IPM in the paddy fields is the integration of multi-resistant cultivars with a corresponding reduced application of pesticides. The introduction of resistant cultivars has also had some success in defending against rice diseases such as blast and brown-spot disease (Mew 1992).

While there has been some success in the implementation of IPM in paddy fields in Malaysia, in general IPM strategies have not been implemented effectively by the paddy farmers (IRRI 1993). The farmers' lack of understanding of pest management concepts has been put forward as one of the contributing factors. Hence, it is recognized that effective IPM programmes must be adapted to the needs of the local farmer. Effective training of the rice farmer has also been identified as an essential component to successful implementation of IPM strategies. Furthermore, maintenance and development of resistance in rice cultivars require continued research encompassing the nature of interactions and the life cycle of the pest (Mew 1992).

Role of NGOs in pesticide governance

There are three non-governmental organizations (NGOs) relevant to pesticide use in Malaysia: the Consumers Association of Penang (CAP), established in 1970, Sahabat Alam Malaysia (Friends of the Earth Malaysia), established in 1980, and the Pesticide Action Network (PAN), established in 1982. PAN currently links over 300 organizations in some 50 countries, coordinated by six regional centres. The registration of these NGOs is governed by the Registrar of Societies Malaysia under the Home Affairs Ministry. The publications of the NGOs are controlled under the Printing and Press Publication Act 1984.

One of the main objectives of these NGOs is to oppose the misuse of pesticides and support reliance on safe and sustainable pest control methods. This is usually carried out by working with communities such as plantation workers, farmers, and smallholders to help them to articulate problems related to their livelihood and living conditions. In addition, NGOs also provide basic consumer education on issues such as food, nutrition, and health. This is done through talks, discussions, house-to-house counselling, slide shows, and exhibitions. Magazines and newsletters such as *Pesticide Monitor* and *Global Pesticide Campaigner* are also produced to share the latest information on pesticides, sustainable agricultural issues, regulatory requirements, and databases, books, and resources. Other areas include research into key issues such as pesticides and credit, national pesticide reduction plans, and advocacy in the regional and international arena. In order to highlight issues on pesticides, NGOs carry out activities such as organizing conferences, training programmes, and talks.

PAN in particular has been paying special attention to persistent organic pollutants (POPs), which include endosulfan and DDT. These pesticides are on the PAN list of chemicals that should be banned from international trade and use due to their adverse impact on human health and the environment. In 1996, PAN Asia Pacific and its network partners took the opportunity afforded by the World Food Summit to launch a safe food campaign advocating healthy eating choices and support for organically grown local food (*Pesticide Monitor* 1997). At the end of 1996, a PAN Asia Pacific video crew flew into the Philippines, Thailand, and Malaysia to obtain footage on pesticide use and abuse and recorded interviews with farmers, plantation workers, citizens' groups, and key scientists. The video, which looked at the impact of pesticides on health, was completed in the early months of 1997.

As a commitment to ensure that consumers have a choice and access to healthy, uncontaminated food, CAP has collaborated with farmers to initiate pesticide-free farming programmes (CAP 1998). Surveys by CAP showed that consumers tend to prefer good-looking, succulent, unblemished produce. However, CAP has launched campaigns to educate consumers that nutrition is more important than appearance while canvassing farmers to decrease their reliance on agrochemicals.

In Malaysia, NGOs play an active role with regard to monitoring and compliance of the national law by sending memoranda to government from time to time through legal advisers (Mageswary 1998). Through the activities carried out by the NGOs in Malaysia, the level of awareness among the community has markedly increased since the 1970s and 1980s. In general, NGOs have received effective cooperation from the relevant government authorities (*ibid.*). Conferences and seminars organized by NGOs have also received a good response from the government. Such events have provided a platform for open discussion with the relevant parties involved. However, NGOs have consistently urged the government authorities involved to be more transparent about their activities and to take into account in a more urgent manner the inputs from the NGOs.

Future outlook for governance of pesticides

Although it is the responsibility of the government to set the standards of plant protection policies, effective implementation is only possible if all stakeholders are committed and involved, especially the private sector. The suppliers are primarily responsible for developing appropriate pesticide products and application equipment, for distributing them effectively, and for making available adequate advice on safe management and effective use of those products and apparatus. Another group sharing the responsibility is the end user. The following are some suggestions as to how the various parties involved can contribute more effectively to minimize the detrimental impact of pesticide use in the country.

How to strengthen the law?

The related laws should ensure that enforcement is effective and make serious provision for the establishment of appropriate educational, advisory, extension, and health-care services. The Pesticides Act should define and clarify the responsibilities of the various parties involved in the development, distribution, and use of pesticides. Predicted impacts of pesticides used at present are largely based on data derived from temperate countries; these data are not always fully reliable. Therefore, there should be a provision to evaluate formulations, storage, and effectiveness under local conditions and to provide appropriate training. Inspectional and laboratory facilities in addition to adequately trained inspectors must be strengthened for effective regulatory control. It is also important that the law allows the industry only to make available less toxic formulations while prohibiting extremely toxic products. In this respect there should also be provisions for the ordinary citizen to participate in a constructive manner. They must have access to information such as health and safety data on chemicals in the market, poisoning records, government enforcement of pesticide regulations, what chemicals are on site at pesticide production facilities, who in the government is charged with receiving notifications of pending shipments of banned pesticides, and who decides to allow or stop their imports.

Role of the government

In addition to the modifications to the Pesticides Act suggested above, the government has a further role to play.

- It should ensure that active ingredients and formulated products for pesticides for which international specifications have been developed conform to the specifications of the FAO and WHO.
- It should keep extension and advisory services as well as farmers' organizations adequately informed about the range of pesticide products available for use in each area.
- It should ensure that pesticide packaging has information and instructions in a form and language adequate to allow safe and effective use. Warning against the reuse of containers and instructions for their safe disposal should be included. Only an adequate quantity should be supplied, packaged and labelled as appropriate for each specific market, including users who are unable to read.
- It should ensure that packaging or repackaging is carried out only on licensed premises where the staff members are adequately protected against toxic hazards, and products are properly packaged and labelled.
- It should assist in the design and development of personal protective equipment more suitable for tropical climates such as Malaysia.
- National or regional poisoning information and control centres should be established to provide immediate guidance on first aid and medical treatment, accessible at all times.
- It should ensure that pesticides available through outlets which also deal in food, medicine, other products for public consumption, or clothing be physically segregated so as to avoid any possibility of contamination or of mistaken identity.
- It should allocate high priority to developing and promoting integrated pest management, and R&D studies should be actively carried out by the relevant authorities to reduce the adverse effects of the development of resistant species.

Role of the pesticides industry

The pesticide manufacturing and distribution industry can be particularly helpful in implementing effective governance of agricultural chemicals.

- It should adhere to the FAO specifications as a standard for the manufacture, distribution, and advertising of pesticides.
- It should pay special attention to formulation, presentation, packaging, and labelling appropriately in local languages (and through pictograms) in order to reduce hazards to users.
- It should take all necessary precautions to protect the health and safety of operative, bystanders, and the environment and at all times should promote the principles as well as ethics expressed by the international code of conduct on the distribution and use of pesticides.
- It should ensure that persons involved in the sale of any pesticides are adequately trained and possess sufficient technical knowledge so that they are capable of providing the buyer with advice on safe and efficient use.
- It should provide a range of pack sizes and types which are appropriate for the needs of small-scale farmers and other local users to avoid handling hazards and risk of products being repackaged into unlabelled or inappropriate containers.
- It should ensure that advertisements do not contain any statement that is likely to mislead the buyer with regard to safety, or the product's nature, composition, or suitability.

Role of the farmers

The farmers, as end users of pesticides and herbicides, have clear responsibilities towards their usage.

- They should always read the label before buying or using pesticides, or have the label read to them if they cannot read.
- They should ensure that all dealers/retailers are adequately trained and possess sufficient technical knowledge to present complete, accurate, and valid information on the product sold, such as recommended rates, frequency of application, and safe pre-harvest intervals.
- They should be aware that any exposure is dangerous and that overexposure may be fatal.
- They should pay attention to the proper use of equipment and the safety measures needed in handling the pesticides. These include the use of mask, gloves, shoes, protective clothing, and mixing apparatus, and special precautions for children and pregnant women. They should also be aware of the danger of reusing containers and the importance of following label directions.
- They should support the concept of IPM as a means to reduce the use of pesticides and insist that government agricultural extension officers give priority to alternative non-chemical pest control techniques.

REFERENCES

Abdul, A. A. R. 1993. "The National Agricultural Policy 1992–2010: Prospects and challenges", *Newsletter of the Centre for Agricultural Policy Studies*, Vol. 8, No. 1, pp. 1–6.

Abdullah, A. R., A. H. Sulaiman, and N. Mohamed. 1994. "Use of acetylcholinesterase activity in fish to monitor aquatic pollution of a rice growing area by organophosphorous insecticides", *Malaysian Applied Biology*, Vol. 22, pp. 129–132.

ADB. 1987. *Handbook on the Use of Pesticide in the Asia-Pacific Region*. Manila: Asian Development Bank.

Bunton, J. 1991. "Pesticides to avert hunger", *Far Eastern Agriculture*, January/February, pp. 8–10.

CAP. 1985. *Pesticides – Problems, Legislation and Consumer Action in the Third World. The Malaysian Experience*. Penang: Consumers Association of Penang, pp. 17–63.

Chee, Y. K., S. A. Lee, A. I. Anwar, L. Teo, G. F. Chung, and H. Khairuddin. 1991. "Crop loss by weeds in Malaysia", in Lee, S. A. and K. F. Kon (eds) *Proceedings of the Third Tropical Weed Science Conference, Kuala Lumpur, Malaysia*. Kuala Lumpur: Perpustakaan, pp. 1–21.

Chin, C. K. 1998. Personal communication.

Claro, C. 1998. "Agriculture – New higher yielding rice", *New Sunday Times*, 30 August, p. 32.

DOA. 1996. *Paddy Statistics of Malaysia*. Kuala Lumper: Department of Agriculture, Peninsular Malaysia, pp. 11–14.

DOA, 2000. *Crop Statistics Report (Laporan Perangkaan Tanaman) 2000*. Kuala Lumpur: Department of Agriculture, Ministry of Agriculture Malaysia.

Ho, N. K. 1992. *Herbicide Application in Rice*. Kedah: Muda Agricultural Development Authority.

Ho, N. K. 1997. *The Role of Extension Program in Integrated Weed Management – With Special Reference to the Muda Irrigation Scheme*. Kedah: Muda Agricultural Development Authority.

Ho, N. K. 1998. "The use of herbicides and herbicide resistance in Malaysia", paper presented at the Meeting and Training on Herbicide Resistance, Taegu, Repubic of Korea, 29 June–3 July.

Ho, N. K., K. Dahuli, Z. M. Zakaria, G. Badron, and G. H. Yeoh. 1996. "Multimedia campaign on weed management in the Muda area", paper presented at Rice IPM Network Conference on Integrating Science and People in Rice Management. Kuala Lumpur, Malaysia, 18–21 November.

Ho, N. K. and Z. Zainuddin. 1995. *2,4-D Usage and Herbicide Resistance Problems in Muda Area*. Kedah: Muda Agricultural Development Authority.

Ibrahim, S. 1992. "Regulatory control of pesticide waste in Malaysia", paper presented at MACA International Pesticide Conference, Kuala Lumpur.

IRRI. 1993. *Pest Ecology and IPM Technology Development*, IRRI programme report for 1993. Laguna, Philippines: International Rice Research Institute.

Ismail, M. 1998. Personal communication from agricultural officer, Communication Division, Department of Agriculture, Malaysia, 17 August.

Jamal, H. H., I. Noor Hassim, H. L. Syarif, S. A. Syed Mohd, and D. Noraziah. 1994. "Health effects related to pesticide use among padi farmers of the Muda area in Kedah", in Nashriyah, M., K. Y. Lum, and S. Ismail (eds) *Proceedings of the Seminar on Impact of Pesticide on the Rice Agroecosystem in the Muda Area*. Bangi: Malaysian Institute for Nuclear Technology Research, pp. 59–69.

Jimenez, B. 1997. "Environmental effects of endocrine disrupters and current methodologies for assessing wildlife health effects", *Trends in Analytical Chemistry*, Vol. 16, pp. 596–605.

Kamaruddin. 1998. Personal communication, assistant director, Soil Science Division, Department of Agriculture.

Kavlock, R. J., G. P. Daston, D. DeRosa, F. Fenner-Crisp, L. E. Gray, S. Kaattari, G. Lucier, M. Luster, M. J. Mac, C. Maczka, R. Miller, J. Moore, R. Roland, G. Scott, D. M. Sheehan, T. Sinks, and H. A. Tilson. 1996. "Research needs for the risk assessment of health and environmental effects of endocrine disruptors: A report of the US EPA sponsored workshop", *Environmental Health Perspective*, No. 104, p. 715.

Kim, K. U. 1996. "Ecological forces influencing weed competition and herbicide resistance", in Naylor, R. (ed.) *Herbicides in Asian Rice: Transition in Weed Management*. Stanford: Institute for International Studies, Stanford University/ Manila: International Rice Research Institute, pp. 129–142.

Mabbett, T. 1991. "Pest management in rice", *Far Eastern Agriculture*, January/ February, p. 9.

Mabbett, T. 1997. "Integrated insect defences", *Far Eastern Agriculture*, January/ February, pp. 9–10.

MACA. 1996. *Annual Report and Directory 1994/95*. Kuala Lumpur: Malaysian Agricultural Chemicals Association.

MACA. 1997. *Annual Report and Directory 1995/96*. Selengor: Malaysian Agricultural Chemicals Association.

Magesway. 1998. Personal communication, scientific officer, Consumers Association of Penang.

Majid, T., B. K. Lim, and A. Booty. 1984. "Implementation of the IPM program for rice in Malaysia", in Lee, B. S., W. H. Loke, and K. L. Heong (eds) *Integrated Pest Management in Malaysia*. Kuala Lumpur: Nan Yang Muda, pp. 319–323.

Malaysian Agricultural Directory and Index (MADI) 1995/96. 1996. Major agricultural commodities. Kuala Lumpur: Agriquest, pp. 175–214.

Malaysian Agricultural Directory and Index (MADI) 1997/98. 1998. Major agricultural commodities. Kuala Lumpur: Agriquest, pp. 107–109.

Meier, P. G., D. C. Fook, and K. F. Lagner. 1983. "Organochlorine pesticide residues in rice paddies in Malaysia, 1981", *Bulletin of Environmental Contamination and Toxicology*, Vol. 30, pp. 351–357.

Mew, T. W. 1992. "Management of rice diseases – A future perspective", in Kadir, A. B. S. A. and H. S. Barlow (eds) *Pest Management and Environment in 2000*. Wallingford: CAB International, pp. 54–67.

Ministry of Health. 1991. *Annual Report*. Kuala Lumpur: Ministry of Health Malaysia.

Ministry of Health. 1994. *Annual Report*. Kuala Lumpur: Ministry of Health Malaysia.

Mohan, V. C. (ed.) 1987. *Pesticide Dilemma in Malaysia. Management of Pest and Pesticide*. London: West View Press, pp. 71–78.

Mokhtar, A. M. 1996. "Safety in the use of pesticides at work", in *Proceedings of International Conference on Pesticides, Kuala Lumpur, Malaysia*, pp. 145–147.

Moulton, T. P. 1973. "The effects of various insecticides especially Thiodan and BHC on fish in paddy fields of West Malaysia", *Malaysian Agricultural Journal*, Vol. 49, pp. 224–253.

Mustafa, A. M. 1997. Annual report to the UNU Project on Environmental Monitoring and Analysis in the East Asian Region on Technology Transfer and Environmental Governance, United Nations University, Tokyo, Japan.

Pesticide Board Malaysia. 1996. *Approved Uses for Commodity Pesticides*. Kuala Lumpur: Pesticide Board Malaysia.

Pesticide Monitor. 1997. "Pesticide Action Network Asia Pacific", *Pesticide Monitor*, Vol. 6, No.1, March.

SAM (Friends of the Earth Malaysia). 1984. *Pesticide Problems in a Developing Country – A Case Study of Malaysia*. Penang: Top Print Enterprise.

Soon, L. G. and O. S. Hock. 1987. "Environmental problems of pesticide usage in Malaysian rice fields: Perception and future consideration", in Mohan, V. C. (ed.) *Pesticide Dilemma in Malaysia. Management of Pest and Pesticide*. London: West View Press, pp. 10–21.

Statistics Department of Malaysia. 1995. *Foreign Trade Statistic 1994/1995*, Vol. 3, Part 1, Section 0-5. Kuala Lumpur: Statistic Department of Malaysia.

Sulaiman, A. H., A. R. Abdullah, and S. K. Ahmad. 1989. "Toxicity of malathion to red tilapia (hybrid *Tilapia mossambica* x *Tilapia nilotica*): Behavioural, histopathological and anticholin esterase studies", *Malaysian Applied Biology*, Vol. 18, pp. 163–170.

Tan, G. H. and K. Vijayaletchumy. 1994a. "Organochlorine pesticide residues levels in peninsular Malaysian rivers", *Bulletin of Environmental Contanmination and Toxicology*, Vol. 53, pp. 351–356.

Tan, G. H. and K. Vijayaletchumy. 1994b. "Determination of organochlorine pesticide residues in river sediments by Soxhlet extraction with hexane-acetone", *Pesticide Science*, Vol. 40, pp. 121–126.

Umakanthan, G. 1983. "Pesticides in our environment – Boon or bane?", in Soon, L. G. and S. Ramasamy (eds) *Proceedings of the Malaysian Plant Protection Society Seminar on Pesticide Application Technology*, Vol. 27. Kuala Lumper: Malaysian Plant Protection Society, pp. 35–39.

UN ESCAP. 1994. *Agropesticide: Properties and Functions in Integrated Crop Protection*. Bangkok: United Nations, pp. 263–291.

Warrell, E. 1997. "Pest control in the paddy fields", *Far Eastern Agriculture*, September/October, pp. 30–32.

Yew, N. C. and K. I. Sudderuddin. 1979. "Effect of methamidophos on the growth rate and esterase activity of the common carp, *Cyprinus carpio*", *Environmental Pollution*, Vol. 18, pp. 213–221.

Yunus, A. and G. S. Lim, 1971. "A problem in the use of insecticides in paddy fields in West Malaysia – A case study", *Malaysian Agricultural Journal*, Vol. 48, pp. 167–178.

3

Chinese perspectives on pesticides in the environment

Quan Hao and Huang Yeru

China's population has doubled during the last four decades, and the food supply has increased by more than two-thirds to meet the demand of this mushrooming population (ADB 1999). This gain has been obtained largely by increasing yields through greater irrigation, increased mechanization, and new, high-yielding varieties of crops. Pesticides have played an important part in achieving this higher yield. As in Malaysia and Thailand, rice is the main crop in China and comprises more than 60 per cent of the food supply (*ibid.*). Therefore, obtaining higher yields of rice has been a national goal for the agricultural sector.

A major portion of the pesticide production in China has been utilized to control pests in rice and to protect rice in storage from fungus. A high demand for pesticides in the agricultural sector has indirectly opened up new opportunities for employment and entrepreneurship. Many factories have been set up to produce pesticides – almost all of these pesticide manufacturing units focus on rice-based products, including insecticides, fumigating reagents, herbicides, and growth regulators. Similarly, a variety of commercial enterprises have been engaged in pesticide supply and marketing, import, and export since the 1950s. The business side of the industry has been equally supported by research institutes and universities. For example, the Shenyang Academy of the Chemical Industry and the Shanghai Pesticide Institute have been performing research and investigation into various aspects of pesticides for years.

As is the case in many other East Asian countries, the use of pesticides has led to a number of problems. Overdosing of pesticides in rice is quite common and the quality of pesticides is difficult to control. Air pollution and water pollution due to improper production and use of pesticides have become more serious in recent years (Hough 1998). In this context, pesticide residues in crops are also getting more attention from consumers and the general public. Governmental institutions and other organizations have done a lot of work at national, provincial, and municipal levels to solve these problems. Non-governmental organizations (NGOs) have also been established to facilitate collaboration between ministries, institutes, and industries. Specific legislation leading to regulations and rules has been enacted to establish standards for production, marketing, and use of pesticides. Similarly, many technical standards and practical methods were set up to stipulate the quality, evaluation, maximum residue limits, and detection of pesticides. This has been matched by an information system to disseminate technical information for new crop varieties, feedback of application conditions, and pollution data.

Range of pesticide-related problems

Misuse of pesticides

The misuse of pesticides has been a common problem in the recent past, mainly because the peasants lacked the appropriate knowledge about pesticides and there were no effective regulations or administrative measures governing their use. The peasants did not know which pesticide was suitable for their crops or exactly how to use it. There were also some inappropriate practices of pesticides application that resulted in harmful residues on the crops. For example, when pests could not be controlled by the pesticide at the recommended dosage, farmers would arbitrarily increase the amount applied. Similar increased applications were made to counter rain after the dosage of pesticide. Cross-pollution of different farmers' pesticide applications in adjacent fields could also increase the actual amount of pesticide applied to an area. Often, the farmers would also use pesticides shortly before harvesting the rice if the pest problems were very serious – again leading to undesirable residue levels. This problem was found to be particularly severe for non-degradable pesticides such as BHC, DDT, and toxaphene, which all persist in the environment.

As a partial solution to this problem, the Chinese government issued a *Good Practices for Pesticide Application* document. This document has been revised three times to fit the changing needs of the agricultural system. As part of the programme, plant protection centres or stations were set up in each county. Many technicians were assigned to villages in each county to guide the peasants in cultivation and correct use of agrochemicals. Publicity through the mass media – including frequently published articles in newspapers and magazines on the use of pesticides – has also helped greatly in this respect. Nowadays the peasants are generally better educated and most of them have some fundamental knowledge of pesticide use. Misuse of pesticides has thus been alleviated to a significant extent.

Variable quality of pesticides

Maintaining a consistent quality of pesticides has been a major issue in China. Because the pesticide supply fell short of demand, several workshops started producing pesticide formulations by mixing openly available chemicals with some additives such as emulsifiers and marketing them to peasants. These workshops – which may have served previously as factories processing chemicals or pharmaceuticals – typically did not have the appropriate certification or skills for making the pesticide products. Consequently, their products would mostly fail to meet the requirements of a pesticide specification or standard. At the same time, the farmers using these "home-made" pesticides might not see the desired results due to this variable quality, and would then increase the dosage. Poor-quality management practices were also present in *bona fide* pesticide manufacturing companies, again leading to overdosage by farmers. In addition to these problems, the impurities in a pesticide often surpassed the regulated quality standard, resulting in poisoning of the user during the application process. The Chinese government has strengthened the administrative oversight of pesticide manufacture during recent years. Those factories without appropriate certification have been banned from pesticide production or processing.

Limited development of pesticide industry

Historically, the pesticide manufacturing industry in China was not well developed. The capabilities of production units and the types of pesticides produced were limited. Distribution of locally produced pesticides was uneven. The factories were mostly only equipped to make chemical pesticides; biological pesticides are still few. Insecticides were the principal products for the Chinese pesticides industry. Low-efficiency pesticides, including some non-prohibited chloropesticides, were used on a large scale. Pesticides with high efficiency and low toxicity were in great demand. This situation resulted in increased imports of unfamiliar pesticides that may have the potential for misuse, overdosage, and new envi-

Table 3.1 Pesticides that have been used for rice

Pesticide category	Generic pesticide names
Organochlorinated pesticides	BHC, DDT, endrin, dieldrin, 2,4,5-T
Organonitrogen pesticides	pyrazoxifen, propiconazole, thiobencarb, bentazone, aldicarb, isoprocarb, oxamyl, carbaryl, fenobucarb, bendiocarb, methiocarb, imazalil, esprocarb, chinomethionate, pirimicarb, flutoluanil, pretilachlor, pendimethalin, metribuzin, mefenacet, mepronil
Organophosphate pesticides	EPN, ediphenphos (EDDP), ethoprophos, etrimfos, chlorpyrifos, dichlorvos, diazinon, thiometon, terbufos, trichlorfon, vamidothion, parathion, parathion-methyl, pirimiphos-methyl, fenitrothion, fenthion, phenthoate, phoxim, malathion
Pyrethroids	cypermethrin, deltamethrin, pyrethrins, permethrin

ronmental problems. Although the Chinese government, factories, institutes, and universities have undertaken actions to change this situation, there is still considerable room for improvement.

There are more than 850 pesticides and formulations registered in China. More than 200 formulations are registered for rice. The main categories and individual pesticides are listed in Table 3.1. Some of these are produced domestically and some of them – especially herbicides – are imported.

Poor control over imported pesticides

Historically, importing pesticides into China was not a well-controlled process. The quality of foreign pesticides was also not always consistent; perhaps as a result of higher-profit motives. In the past China imported several million dollars worth of pesticides each year – Table 3.2 provides a partial list of the international corporations and their products sold in China. On a national scale, pesticides were imported without a coordinated plan. This also indirectly affected the plans for production, delivery, and use of domestic pesticides. In addition, it was more difficult to ensure that foreign pesticides had been used correctly.

In 1982 six ministries and agencies – Agriculture, Public Health, Chemical Industries, Trade, Forestry, and the National Environmental

Table 3.2 Partial list of companies and pesticides registered in China

Company	Registered pesticides
Bask Aktiengesellschaft	Basagran, Blazor, Calixin
Bayer	Hinosan, Bayleton
Ciba-Geigy	Ridomil, Dual
Dainippon Ink and Chemicals	Befran
Ishihara Sangyo Kaisha	Fluazifop-butyl
Kumiai Chemical Industry Co.	Saturn, Basitac
Mitsubishi Kasei Corporation	Mipcin, Bassa, Marvric, Nomolt
Nippon Kayaku Co.	Diazinon
Nippon Soda Co.	Topsin-M, Nissorun, Napu
Sumitomo Chemical Co.	Sumicidin, Sumi-alpha, Sumithion, Meothrin, Sumilex
Uniroyal Chemical Company	Comite, Vitavax
Shell China	Ripcord, Fastac
Schering	Mitac, Betanal AM11
Sandoz	Ekallux, Evisect, Marvric, Sandofan, Banvel
Roussel-Uclaf	Decis, K-othrin, K-obiol
Rhone-Poulenc Agrochimie	Temic
Imperial Chemical Industries	Primor, Cymbush, Kungfu
FMC Corporation Agricultural Chemical Group	Furandan, Arrivo
Duphar B.V. Weesp, Holland	Dimilin
Dow Chemical Pacific	Lorsban, Gallat, Starane

Protection Agency – issued cooperatively the Regulations for Pesticides Registration. Since then, foreign pesticides cannot be imported and marketed in China without approval and registration in accordance with these regulations. Imported pesticides must be inspected by the Administration of Commodity Inspection (now the State Administration for Entry/Exit Inspection and Quarantine).

Environmental pollution caused by pesticides

There is an overall problem of pollution to ecosystems and the environment caused by pesticides. The pollution problem has received a lot of public attention. It is apparent that the regulations governing the research, production, transportation, supply, sale, and application of pesticides need to be effectively applied. There is a need for a monitoring system to evaluate the effectiveness of the regulations. There are numerous routes and ways in which pesticides may pollute the ecological system at different stages of manufacturing, transportation, storage, supply, sale, and application.

- *Pollution of agricultural products such as rice.* This may be caused by overdosage, short period of dosage before harvest (short withdrawal time), or drift of pesticide granules or dips to the adjacent area during application.

- *Water source pollution.* Pesticides used in the field are washed or directly carried to the next field or underground water. Once the surface water been polluted, it will further pollute the aquatic organisms. Pesticides in underground water can persist for a long time due to the poor capacity for self-purification of the groundwater.
- *Air pollution.* This mostly happens during pesticide application, especially in plane puffing. The drifting aerosols pollute the adjacent areas. The air around pesticide production and storage sites may also be easily polluted.
- *Soil pollution.* When the residual pesticide concentration is above a certain level, it may spread to the surrounding area and endanger the adjacent planted crops.
- *Ecosystem pollution.* Some high-residue pesticides may affect the whole ecological system through bioaccumulation in the food chain. Common examples of this are the organochlorinated pesticides, which have been shown to persist and bioaccumulate in ecosystems.

A new generation of pesticide substitutes for current pesticides also needs to be developed. This requires a long-term plan for the development of new techniques for pest control, the decrease of pesticide use, and the prediction and observing of pollution and protection of the environment.

The best way to alleviate pesticide pollution is to reduce the use of pesticides in the first place. In 1983 the Chinese government decided to ban the production of BHC and DDT, and gradually prohibit the use of those two pesticides in agricultural products. Some extremely toxic pesticides were regulated such that they can be used in some special cases. The State Scientific and Technological Commission has supported several research projects for the development of low-toxicity and high-efficiency pesticides or biological pesticides. A programme called the Green Food Project was initiated by the Ministry of Agriculture. Under this programme, crops were cultivated without the use of chemicals in a special area. Many food products without residues can be found in the market. This programme has also encouraged research and development for new varieties of crops with resistance to some insects.

Monitoring of pesticide residues

At present there is no comprehensive system for the monitoring of pesticide residues in rice or other crops for domestic consumption in China. But the research institutes and sanitary and quarantine departments are the main players. It is believed that the sanitary quarantine stations at county, provincial, or municipal levels under the Ministry of Public Health have performed satisfactorily for many years. These stations conduct the sampling themselves, analyse the residue, and make a report to the relevant departments. The imported (mainly wheat) and exported (mainly rice) crops are tested by the State Administration for Entry/Exit Inspection and Quarantine. The pesticides that have been tested in recent years are listed in Table 3.1. According to the monitoring report, no problem cases have been detected during this monitoring – in other words, there have been no residue data that surpass the corresponding maximum residue limits. Very low levels of a few pesticides have been found in unpolished rice and imported cereals. For example, some fumigating reagents and chlorinated pesticides were found at levels below the limits in domestic rice and imported wheat. The China-Japan Friendship Center for Environmental Protection has undertaken a project for environmental monitoring and analysis in which polished rice samples were collected from different areas and analysed by GC/MS. No significant level of pesticides was detected (Hao 1996).

Governance of pesticides

Pesticides are unique as a product in that they affect agricultural production while being hazardous chemicals to manufacture and market, and toxic to the human population. Therefore, governance of pesticides requires a series of laws, regulations, and standards concerning pesticides administration, pesticides quality, safety in pesticides production, and pesticides inspection.

Legislative framework in China

Legislative development for pesticide-related issues is a continuous process: some related laws or regulations have already been issued, and still more are being drafted. The Environmental Protection Law, promulgated in 1979 and reworked in 1989, regulated the overall policy and principles for environmental protection. It is a highly complex set of regulations and is the mainstay of the governance infrastructure for environmental protection. A law on the control of toxic materials and a law for protection of the agricultural environment and the registration of toxic chemicals are already drafted. Regulations for food hygiene have been issued and promulgated by the Ministry of Public Health; specific articles pertaining to food safety also include provisions for pesticide residues.

In 1984, an announcement on production, processing, supplying, and marketing of rodentists was jointly published by the Ministries of Trade, the Chemical Industry, Forestry, Transportation, Public Health, Agriculture, and some other administrative units. A circular on the "Prohibition of Production, Import, Marketing, and Use of Ethylene Bromide" was issued in December 1984. Another circular was issued by the Ministries of the Chemical Industry, Trade, and Agriculture and the State Planning Commission in February 1990 to cease the production of chlordimeform within three years. The Department of Chemical Industries was instructed to make a plan to eliminate the production of chlordimeform and to identify a substitute for it. Departments of Agricultural Plant Protection, Trade, and Supply and Marketing were appointed to guide pesticide usage and assure that application is in accordance with the requirements of *Guidelines for Safe Application of Pesticides* (State Planning Commission 1989). In addition, a supplement was published by the Ministry of the Chemical Industry to regulate foreign investments in the pesticide industry, construction of new plants for pesticide manufacture, the direction of new pesticide development for restricted, prohibited, and banned pesticides, and certification for pesticide production. In 1992, the State Council developed a circular on strengthening the administration for pesticides and veterinary drugs, issued by the Ministry of the Chemical Industry. It was emphasized that the production and use of BHC, DDT, dibromochloropropane, and chlordimeform are prohibited.

A set of Agrochemical Products Administrative Protection Regulations were approved by the State Council and promulgated by the Ministry of the Chemical Industry in 1993. The general purpose of these regulations, and specifications contained therein, is to protect the rights of overseas manufacturers of agrochemical products operating in China. The requirements and procedures for application, assessment, and certification are specified in some detail.

It is important to note that pesticide registration is at the core of pesticide administration. This is a system governing the activities concerning pesticides before marketing. All factories must apply to register their products, including supplying technical details and formulation, and obtain approval from the government for certification before production and marketing.

In April 1982 the Pesticide Registration Regulation was issued by the Ministry of Agriculture in collaboration with the Ministries of Forestry, the Chemical Industry, Public Health, and Commerce and the Leading Group of the State Council for Environmental Protection. It was decreed that domestically developed new pesticides must be registered before production. Any pesticides without such approval should not be produced, marketed, or used. Old pesticides developed before the regulation was issued had to be re-evaluated and re-registered. Foreign pesticides could not be imported without prior approval. Since then 16 specific provisions have been issued to modify the regulation.

The 1982 regulation applies to registration of a broad range of chemicals and biological products. These include all pesticide and herbicide products used for the protection of agricultural, forestry, and animal husbandry crops and spices from attack or effects of pests, insects, fungus, weeds, or any other organisms (including vector control for public health reasons). Registration comprises three steps: field testing, tentative registration, and formal registration. Field testing is a period for the pesticide to be examined for its quality. An application for registration in China should be linked to information such as pharmaceutical and residue data obtained from China. Two-year tests must be carried out in two areas of China with different natural conditions to obtain these data. Tentative registration is needed while a model test is carried out on a one-acre or three-acre area, or for a trial sale period. A pesticide circulated in the market should be fully registered. The validation time is five years. A registered pesticide should supplement its registration description when its content or formulation is changed.

National standards in China

There are more than 40 national standards and 85 special standards for the specification of pesticides. Pesticide quality is specified by one of these national or ministerial standards. The specifications include the content of the active component in a technical pesticide or formulation, and the chemical and physical properties. The requirements of agricultural application and factory production are also considered when drafting or adapting a standard.

There is a supervision system in China to oversee the quality of registered pesticides – the State Administration for Technical and Scientific Products is responsible for the supervision of pesticides. Registered pesticides are checked while they are in circulation to assess whether their quality is in accordance with the requirements of the promulgated standards. Uncertified products are prohibited from entering the market to protect the interests of consumers and agriculture. Under the direction of the State Administration, two inspection centres were set up in Beijing and Shenyang for the supervision and inspection of pesticides. The main tasks of the two centres are to:

- carry out supervision inspections
- test for pesticide quality
- investigate, draft, and adapt pesticide standards
- develop new methods and technologies for pesticide analysis
- carry out pre-production evaluation for a new pesticide

- organize technical exchanges domestically and internationally
- provide training to technical workers in pesticide analysis for factories and regional administrative organs.

The provincial or municipal supervision organs set up pesticide supervision stations to manage the inspection for a specific area. Inspections for pesticides quality are made periodically or randomly. Uncertified products and producers are publicized all over the country. Additionally, producers found to be responsible for the production of sub-standard pesticides are not allowed to make pesticides until their products have been determined to show a good quality.

Pesticide residue can be defined as the total toxic compounds of the parent pesticide: the metabolites, decomposition products, and impurities that persist in biological tissues, agricultural products, and the environment after its use. If the amount of the residue surpasses the maximum residue limit (MRL), it can be dangerous to human health as well as to farm animals and other fauna and flora in the ecological system. The Ministry of Public Health has issued and promulgated several national standards for pesticide residues. The published data are listed in Table 3.3. All residues are given for unhusked raw cereal except dimehypo and thiocyclam, which are for husked rice. More national standards for more new pesticides are still under development. It is important to note that since 1983 China has gradually prohibited the production and application of hexachlorocyclohexane and DDT.

All the MRLs set by the Ministry of Public Health are accepted by all ministries in China. These data are set up in accordance with the principles laid down by the Codex Committee for Pesticide Residues (CCPR) of the FAO. Because the Chinese dietary model is different from other countries – rice consumption is higher in China than in European countries – lower maximum residue limits may be set for some pesticides in rice. While some MRLs are given for unhusked rice, these may be higher than those for husked rice in some countries, for instance Japan.

Information management and dissemination systems in China

The IRPTC-NRPTC system

In 1976 a system called the International Register of Potentially Toxic Chemicals (IRPTC) was developed in Geneva. It was designed to comprise a network for data collection and a central unit. The IRPTC network partners consist of national correspondents, national and international institutions, industries, and external contractors. China became

Table 3.3 Maximum residual limits of pesticides in rice

Pesticide	Allowance (ppm)	Pesticide	Allowance (ppm)
Acephate	0.20	Aldrin	0.02
Arsenicals	0.70	BHC	0.30
Carbon disulphide	10.00	Chloropicrin	2.00
Chlorpyrifos-methyl	5.00	Cyanide	5.00
DDT	0.20	1,2-dibromoethane	0.01
Dichlorvos	0.10	Dieldrin	0.02
Dimethoate	0.05	Fenitrothion	5.00
Fenthion	0.05	Heptachlor	0.02
Malathion	3.00	Mercury	0.02
Parathion	0.10	Phorate	0.02
Phosphide	0.05	Anilazine	0.20
Buprofezin	0.30	Carbaryl	5.00
Carbendazim	0.50	Carbofuran	0.50
Chlormequat	5.00	Chlorothalonil	0.20
2,4-D	0.50	Deltamethrin	0.50
Diazinon	0.10	Ethion	0.20
Fenvalerate	0.20	Flucythrinate	0.20
Methamidophos	0.10	Methyl bromide	50.00
Parathion-methyl	0.10	Permethrin	1.00
Phosmet	0.50	Phosphamidon	0.10
Phoxim	0.05	Pirimicarb	0.05
Pirimiphos-methyl	5.00	Propiconazole	0.10
Quitozene	0.10	Thiocyclam	0.20
Triadimefon	0.50	Triadimenol	0.10
Trichlorphon	0.10	Tricyclazole	2.00
Isocarbophos	0.10	Dimehypo	0.20

a full partner of the IRPTC network when the Chinese government in 1979 decided to appoint the Institute of Environmental Health Monitoring of Chinese Preventive Medicine as the national correspondent for IRPTC in China.

In 1986, the IRPTC-NRPTC concept was introduced and the National Protection Agency was appointed as the organization responsible for the set-up of the National Register of Potentially Toxic Chemicals (NRPTC). In 1988 a workshop was held in Beijing in collaboration with UNEP to establish a national information system on potentially toxic chemicals. Since then substantial progress has been made on the development of the NRPTC. Indeed, a national platform for discussion on chemicals is being developed, a national network has been convened, and communication channels for the transfer of information amongst the network are being developed. A start has been made with national data and storage in formatted worksheets or computerized data files compatible with the IRPTC register.

The information on chemical compounds registered in the NRPTC includes:

- identifiers and properties such as chemical and common names, trade names, CAS number, molecular formulas and weight, definition, additives, and impurities;
- production and trade, including dates, producer's name, importer's name, formulator's name, formulations with percentage of active ingredient, means of distribution, and quantities;
- production and formulation processes and impurities associated with these processes;
- uses and date, quantities, and geographical area;
- loss, persistence, residue studies, concentrations, and possible human intake;
- environmental fate tests and pathways into the environment;
- chemobiokinetics;
- mammalian toxicity and special toxicity studies, including effects on organisms in the environment;
- spills and their location, monitoring of concentrations after spills, and remediation measures taken;
- treatment for poisoning;
- waste management;
- recommendations/legal mechanisms.

Most pesticides belong to the category of toxic chemicals and are therefore contained in the target compound series of the IRPTC information system. Registration of agrochemicals has been formed as a regulatory procedure in China. As discussed earlier, several specific rules for safe use, classification, application, transportation, and storage have been issued.

Within this context it is important to mention the Chinese policy for the management of toxic substances, as recently formulated by the Environmental Commission under the State Council. This commission, in which all the agencies involved in environmental affairs are represented, agreed on the following points:

- setting up of one organization through which all matters related to the registration of chemicals, priority setting of chemicals, risk/hazard assessment of chemicals, and other matters will be coordinated;
- the importance of letting specific ministries (Public Health, Chemical Industries, Agriculture, Labour) maintain their responsibility for chemical management in their specific field;
- the appointment of the National Environment Protection Agency

(NEPA) for the coordination of all activities in the field of chemical management and regulations.

In order to initiate a structure for a national network of selected suppliers and users of data on potentially toxic chemicals, the situation in China has been organized as follows.

- Institutes at all organization levels (central, provincial, and county levels) have been given a mandate, pursuant to the Environmental Protection Law (1979), to perform hazard/risk assessments.
- Under the guidance of the NEPA, the people's governments of the provinces, autonomous regions, and municipalities directly under the central government (Shanghai, Beijing, Tianjin, and Chongqing) have established environmental protection bureaus (PEPBs).
- Recently PEPBs have started storing data available at provincial level on chemicals in computerized systems.
- At present, exchange of information is possible between national and provincial data systems but compatibility problems still hamper free communication, even though the internet has been used in China for a few years in some fields.
- The NEPA will not be able to reach all potential provincial network partners, so delegation of responsibilities is necessary and important.

Bearing the two latter points in mind, the provincial network partners were informed on the contents, structure, and concepts of the IRPTC-NRPTC programme. In this respect, provincial partners will have a delegated responsibility in data storage, retrieval, and dissemination and the establishment of Provincial Registers of Potentially Toxic Chemicals (PRPTCs).

Representatives of provincial bureaus involved in the management of chemicals have expressed keen interest in becoming full network partners, through which they will be better equipped for performing hazard/risk assessments in their own region. Moreover, their participation in the network may start to improve the coordination between network partners at provincial levels, between representatives from the public health sector, the industrial sector, the labour sector, the agricultural sector, etc. At present, the lack of coordination between these provincial network partners is considered to be of one of the more important reasons for insufficient control of chemicals.

Establishment of ICAMA

To implement the Regulations on Pesticides Registration, the Institute for the Control of Agrochemicals of the Ministry of Agriculture (ICAMA) was appointed as the responsible body. The main purpose of these regulations, as explained earlier, was that pesticides without registration should not be distributed in the market. Based on the information generated by applications under the regulations, a central filing system containing relevant information on industrial patterns, trade flows, and health and environmental hazards posed by pesticides has been constantly given priority attention by the ICAMA. Active participation of network partners has been sought and several mechanisms for collaboration and various levels of assistance to registration and information provision have been elaborated.

The data profiles enable the expert user to identify what is known about a particular pesticide in terms of its chemical, physical, environmental, and toxicological characteristics (ICAMA 1992). In addition, they provide information on production and consumption, use, spills, treatment of poisoning, and waste management, as well as recommendations and legal mechanisms for control of hazards posed by pesticides.

The institute has assisted the local governments of provinces, counties, and districts in setting up control departments for agrochemicals, training technical personnel, and giving direction in technical works. It has also taken part in several important international meetings on pesticide administration. It became the Chinese appointed coordinator in the network for coordination in the Asia Pacific area, and in the harmonization network of the International Code of Conduct on the Distribution and Use of Pesticides, organized by the FAO.

Publications on pesticides

A bulletin for registered pesticides is published monthly, and a cumulative report on registered pesticides is published every year (ICAMA 1989–1995). It contains information on factories or corporations which apply for registration, the scope of application for the registered pesticide, and its formulations. In addition, information on physical, chemical, and toxicological characteristics, effects and dynamics, environmental ecology, labels, and instructions for use can be obtained by searching a computerized file or the library of the Agricultural Sciences Academy.

There are a wide variety of books, handbooks, encyclopedias, and magazines available that provide pesticide-related information. Chinese magazines such as *Pesticides* (edited and published by Shinnying Institute of Chemical Industries), *Translated Collection of Pesticides* (edited by the Shanghai Pesticide Institute), *Pesticide Science and Administration* (published by ICAMA), and *Pesticide World* can be found in every agricultural-related library. An encyclopedia for agriculture in China has been published in Chinese and there is a separate volume for pesticides. An expert user can easily obtain information on the characteristics, use, and environmental hazards of a pesticide.

Although the governmental network needs more time to update its computer system, non-governmental networks could be used in information exchange. The internet is now very popular in universities, institutes, and some families. Most of the relevant governmental organs, institutes, factories, corporations, and even personnel have their own home pages and/or e-mail address. It is therefore easy for an expert to get information on pesticide characteristics, agrochemical industries, statutes related to pesticide application, international guidance, and regulations and conditions governing residue in crops and food products. Some libraries provide services to people who do not have a computer (or registered internet address) for literature searching on the internet. The information system of the Association of Libraries, the information system of the Academy of Sciences, and the Chinacom system can be used for searching for pesticides information published in Chinese.

There are several pesticide research institutes that belong to the Ministry of the Chemical Industry and Engineering, the Ministry of Agriculture, the Academia Sinica, and the higher education system. The institutes under the Ministry of the Chemical Industry and Engineering include the Shenyang Academy of Chemical Engineering, Shanghai Pesticide Institute, Jiangsu Pesticide Institute, Anhui Institute of Chemical Industry, Zhejiang Academy of Chemical Engineering, and Hunan Academy of Chemical Engineering. These institutes mainly conduct research on pure pesticides and synthesis of pesticide delivery mediums, new pesticide development, and industrialization of new pesticides and formulations. The Shanghai Institute is the main body studying antibiotic pesticides. The Jiangsu Institute is the first unit in China to develop pyrethroids. The Hunan Academy focuses on carbamates. Liaoning Academy of Agricultural Sciences focuses on rodentists. The pesticide branch of the Plant Protection Institute of the Chinese Academy of Agricultural Sciences has studied organochlorine and organophosphorus pesticides for about 40 years. It has been engaged in the study of the basic theory, technology, and application of pesticide usage. Beijing Botany Institute under the Academia Sinica carries out the development of plant growth regulators.

The role of non-governmental organizations

There are several non-governmental organizations with activities related to pesticides. They were set up under the guidance of various ministries, including the Ministries of Agriculture, the Chemical Industry, Public Health, Trade, and Forestry, and the NEPA, separately or cooperatively. These organizations have played an important role in organizing academic activities and training courses, carrying out investigations, and presenting recommendations based on scientific factors for government decisions in formulating policies, strategies, and plans for pesticides. They have acted as the bridge and link between governments and industries. Some of the organizations are described in this section.

China Pesticide Industry Organization

The China Pesticide Industry Organization comprises pesticide industries and institutes. It was established in Beijing in 1982 for the harmonization of and service to economic and technological affairs in pesticide industries. It acts as a bridge between governments and industries. There are 145 member units in the organization; between them they have a 90 per cent share of the total output, gross value, and interests and taxes in pesticide enterprises. The main tasks of the organization are as follows.

- *Investigation* – for example, research on pesticide pricing, processing technologies, advances in factory processes, and administrative issues for the industry. This provides a scientific basis to governments for formulating pesticide plans and policy administration, and enforcing the administration of the pesticide industry.
- *Organization* – this includes labour emulation drive, exchanging experience, and promoting technical competence and the economic benefits of pesticides.
- *Publishing* – the organization publishes special journals, and transmits messages concerning production, scientific research, and application.
- *Training* – a training centre was set up in the Hangzhou pesticide factory in 1989; since then training courses have been held on industry management, flow meters and measuring meters, application of pesticide emulsifiers, etc.
- *Establish a foundation* – in order to encourage the growth of knowledge in the industry, the association has set up a scholarship programme. This programme was established to recognize and reward excellent graduates and undergraduates in Beijing Agricultural University, Nankai University, Middle China Normal University, Zhejiang Agricultural University, etc.

The Chinese Society of Pesticide Science

The CSPS is a branch of the Chinese Chemical Industries Engineering Association. Its main task is to organize academic activities for the exchange of pesticide science achievements, study the strategy for the development of pesticide science and technology, and promote the development of pesticide industries. It was founded in Shenyang in 1979.

The society has taken an important part in promoting the development of Chinese pesticide science and industries and the domestic and international exchange in pesticide technology. Since 1979 the society has held meetings every other year to discuss special topics and exchange dozens of research papers. Besides these meetings, the society has organized several symposia and seminars on the development of pesticide processing technology and the development of new pesticides. The society has held special topic discussions for the national pesticide development plan and has presented particular recommendations to the sixth five-year plan and subsequently. The suggestions provided by the CSPS for the development of pesticides in 2000 and for achieving energy efficiency and decreased consumption were adopted by some administrative departments. Its proposal for the invention of new pesticides was confirmed by the State Scientific and Technological Commission and the State Educational Commission. A project for computer-assisted pesticide molecular design was supported by the State Planning Commission.

It is important to note that exchanges have been held several times between the mainland and Taiwan since the first symposium in Shanghai initiated by the president of the society, Zicheng Xu, and Maokun Shi in Taiwan. This has strengthened mutual understanding and cooperation between the two sides of the Taiwan Straits and has promoted trade.

A Sino-Japanese Mutual Pesticide Societies Symposium was initiated by the president of the society in 1982. Since then it has been held every other year alternately in China and Japan; it has been held several times separately in Hangzhou, Tokyo, Beijing, Kyoto, Xi'an, etc.

Assessment Committee for Pesticide Registration

The Assessment Committee for Pesticide Registration is an expert organization for providing assistance in pesticide registration; it was founded in 1982, and is under the charge of the Ministry of Agriculture. Its main tasks are to make a comprehensive assessment of pesticides that have formally applied for registration, and to make recommendations on principles and policies for Chinese pesticide administration. Committee members are designated by the Ministries of Agriculture, the Chemical Industry, Public Health, Trade, and Forestry, and the NEPA. Some experts in pesticide technology and administration from the Academia Sinica and other universities are invited. The total number of members in a session is 37, each serving for a term of three years. Five subcommittees have been established for hygienic toxicology, environment, production, circulation, and efficiency and residues. A plenary session and some special meetings are held every year.

Role of the pesticide industry

Magnitude of the industry

China is the third largest pesticide-producing country in the world. The industries made 148 kinds of technical pesticide products in 1990, including 70 insecticides, 40 fungicides, 29 herbicides, and nine plant growth regulators. Insecticides are still by far the main pesticide products, with total production in excess of 170,000 tonnes annually. More than 700 units, including chemical and pharmaceutical factories all over the country, have been approved to produce pesticide chemicals and formulations. More than 200 of these industries operate at a national scale. Every province, autonomous region, or municipality directly under the central government has its own pesticide factories, but most of the factories are concentrated in the agricultural and industrial relatively developed regions such as the coastal, north-eastern, northern, and middle-south parts of China. There are 12 main pesticide factories distributed in Tianjin, Hunan, Shanghai, Jiangsu (Suzhou and Nantong), Hangzhou, Shandong, Qingdao, Zhengzhou, Shanxi, Guangzhou, and Chongqing, employing more than 20,000 people.

Jiangsu province has had almost 200 plants and institutes engaged in the production or development of pesticides, including two provincial-level research institutes devoted to the development of new pesticides and higher educational institutes that carry out research and toxicity tests on pesticides. A complete pesticide industrial complex has been established, ranging from research institutes, higher educational institutes, and agriculture science systems to pesticides enterprises which are dedicated to research, development, toxic testing, popularization of new products, experimental demonstration, and pesticide production.

Many foreign companies have registered their products in China (see Table 3.2); some of these are registered for rice, some for fruit trees, some for cotton, and some for use by smallholders. There are also many related chemical and pharmaceutical factories, such as the eighteenth pharmaceutical factory in Shanghai, Tianjin Dagu chemical plant, factories of the Hormone Institute in Jiangsu province, Yixin biochemical plant, Xuzhou biochemical plant, Jiangxi biological pharmaceutical plant, and Jiangxi pharmaceutical plant.

Each pesticide factory plays an important role in the local area. Generally, it provides most of the pesticides used for local consumption. Besides the agricultural products provided by pesticide plants to the local area, these units have also generated ample opportunities for employment. Pesticide manufacturing units and related industries also contribute to local government through a significant contribution of taxes. In certain rural regions, pesticides plants are the largest taxpayer. On the other hand, it is not unusual to hear complaints from the localities around the plants. The waste from the plants can potentially pollute the air, the soil, and the water – affecting the everyday life of the people living beside the plant site.

Role played by the industry

Since the third plenary session of the Communist Party of China (CPC) Thirteenth Central Committee, changes from a planned economy to a market economy have been taking place all the time. The government macroscopically adjusts and controls the production, research direction, supply and marketing, and application of pesticides by regulations and guidance. However, the industry is increasingly self-regulating and dealing with issues on its own. For example, it has conducted the following types of activities:

- financing the development of new pesticides
- conducting bioassays such as field trials of pesticides
- studying the toxicity and toxicology of pesticides
- monitoring the residue level and analytical method of pesticides
- investigating the best methods of application.

The pesticide industry has also performed significantly in providing technical support for pesticide use – establishing application methods and prepared instructions for customers. Sometimes training courses are necessary for a new type of pesticide, and these are developed and delivered by the industrial sector. Although there are many plant protection centres or stations at county level and technical personnel are available in every village, there is still a greater need for improving the response to complaints and feedback of information on efficiency and convenience in use. It is no surprise that amongst the industry, competition is based in part on the assistance level provided to consumers.

Role of the general public

The general public has been paying much greater attention to environmental issues and their linkage to pesticide production and application. This has matched a generally increased awareness of environmental pollution problems. Most people now know the types of pollution – aerosol, flying dust, automobile exhaust, "white pollution", toxic chemicals, etc. – because newspapers, electronic media, and the television report on these quite frequently.

This heightened public awareness and attention to pesticide-related problems was initiated by the announcement of the State Council in 1983 prohibiting the production, supply and marketing, and use of BHC and DDT. These pesticides had been used widely in China for many years and were very popular in agriculture and the pesticide industry. Almost all crops used to be treated with BHC and DDT for pest control. Many manufacturing plants were solely devoted to the production of these two pesticides. Then knowledge of the problems caused by these pesticides' residues was made public – for example, residual BHC and DDT can lead to cancer and the birth of deformed children.

The public, in return, paid more attention to this issue and responded by reporting more cases concerning pesticides. Farmers noted that pesticides washed by water into fish pools led to the death of fish. Similarly, people living around pesticide manufacturing plant sites have been reporting the pollution caused to their agricultural environment by wastes from factories for many years. Any accidents in production, transportation, supplying, and application were rapidly reported. On the whole information exchange was greatly enhanced as a result of increased awareness. The circular for the prohibition of DDT and BHC may have been based originally on the recommendation of experts on pesticides. They knew the story of *Silent Spring* (Carson 1962) – which marked the beginning of environmental protection – and the toxicological results of pesticides. Mass media have also acted accordingly – numerous articles have been published in newspapers and magazines describing the pesticides' properties, their effects on the environment, and their fate in the environment.

REFERENCES

ADB. 1999. *Key Indicators of Developing Asian and Pacific Countries*. Hong Kong: Oxford University Press.

Carson, R. 1962. *Silent Spring*. New York: Houghton Mifflin.

Hao, Q. 1996. *Annual Report to the UNU Project on Environmental Monitoring and Analysis in the East Asian Region on Technology Transfer and Environmental Governance*. Tokyo: United Nations University.

Hough, P. 1998. *The Global Politics of Pesticides – Forging Consensus from Conflicting Interests*. London: Earthscan Publications.

ICAMA. 1989–1995. *Annually Cumulative Bulletin for Registered Pesticides*. Beijing: Agricultural Publishers.

ICAMA. 1992. *Pesticide Science and Administration*. Beijing: Institute for the Control of Agrochemicals of the Ministry of Agriculture.

State Planning Commission. 1989. *Guidelines for Safe Application of Pesticides*, GB8321.3-89. Beijing: State Planning Commission.

4

Thailand's perspectives on pesticides in the environment

Monthip Sriratana Tabucanon

The Kingdom of Thailand occupies close to 500,000 square kilometres with a population of approximately 60 million. Thailand's agricultural sector is the foundation of the country's economy, with about 70 per cent of its working population engaged in this sector. With the rapid expansion of the overall economy in Thailand, there have been some structural adjustments in the agricultural sector and a diversification of agricultural production. Nevertheless, the largest share of agricultural production is still occupied by rice production and Thailand remains one of the world's leading rice exporters. The demand for agricultural productivity as well as the expansion of industry and the shrinkage of land available for conversion to farming has led to a rapid increase in the use of agrochemicals.

In the pattern of pesticide use in Thailand, rice continues to be the major crop. Rice is grown on around 44 per cent of the total agricultural land and has a 20 per cent share of the pesticide market. Farmers use pesticides intensively – there is little control of the amount used and the application frequency. Often the farmers' lack of awareness is seen as one major reason for pesticide problems. With the increasing trend of pesticide use and the continuing insufficient efforts to control pesticide hazards, the negative consequences will become more apparent in the future. Pesticides are recognized as a major group of imported chemicals from developed or industrialized countries. Total consumption of pesticides in Thailand grew from about 13,000 tonnes in 1985 to over 25,000 tonnes in 1996, as shown in Table 4.1.

Table 4.1 Quantity of imported pesticides in tonnes (1985–1996)

Year	Insecticides	Fungicides	Herbicides	Others	Total
1985	5,146	2,646	4,830	210	12,832
1986	5,799	2,512	4,262	204	12,777
1987	5,881	4,530	3,967	247	14,625
1988	7,050	4,362	5,596	205	17,213
1989	6,937	4,724	6,747	317	18,725
1990	7,176	2,800	8,272	346	18,594
1991	5,560	2,087	7,071	311	15,029
1992	6,098	3,513	8,450	418	18,479
1993	5,305	3,988	9,056	476	18,825
1994	5,252	4,885	9,554	640	20,331
1995	6,573	4,828	11,934	727	24,062
1996	6,608	4,446	14,041	446	25,541

Industries related to organic and inorganic chemicals such as petrochemicals, electroplating, surfactants, plastics, agrochemicals, dyes, drugs, and intermediates have been developed to levels at which they make a significant contribution to the national economy. At present there are more than 70,000 registered factories around the country, of which up to 40,000 are reported as polluting and producing hazardous wastes. During the period 1982–1992, the annual growth rate of the agrochemical market in Thailand amounted to 8.8 per cent. The rate of increase has slowed down in recent years. For example, the expansion of the agrochemical market was predicted to slow down to an annual growth rate of 2.5 per cent for the 1993–1998 period (Wood Mackenzie & Co. 1993). In 1994, the pesticide market reached a sales volume of US$247 million (Mills 1994).

Key issues

The pesticide market and industry

Thailand's pesticide market is handled by the private sector. Imports can be either formulated products or active ingredients which are formulated in the country. The latter sector has an increasing import share, and many international pesticide companies have established formulating plants in the country or cooperate with local manufacturing facilities. The only pesticide manufactured in the country is paraquat: two manufacturing plants produce 5,500 tonnes per year (Tayaputch 1992). For the herbicide market, plantation crops and the rice sector are of great importance. Most important for the insecticide market are the rice and horticultural sectors. The horticultural sector is most important for the fungicide market.

Distribution of agrochemical products in Thailand is usually a two-stage process. Products are sold from the producer or formulator to dealers, and afterwards to subdealers or retailers. The pesticide companies employ sales personnel for the wholesale business as well as for retail at the farmers' level.

One problem associated with pesticide production is the insufficient quality of the products and labelling. In a survey of 373 randomly selected pesticide formulations conducted in 1983, 44 per cent of the samples differed significantly from the indication on the label (*ibid.*). Taking this into account, farmers have limited opportunities to control the amount of active ingredients sprayed on their land. Several studies indicate that farmers found from experience that the amount recommended on the label was not effective and consequently started to apply higher quantities (Grandstaff 1992).

Environmental problems

Recently, surveys and studies on chemical residues in the environment have been undertaken by several institutes. The situation regarding chemical residues in the environment seems to be unsatisfactory. Many toxic chemical residues such as heavy metals and pesticides have contaminated the environment. During the 1980s and 1990s many regions of the country, particularly in the major river basins such as the Chao Phraya and Thachin rivers, have been polluted. The Department of Agriculture reported that three groups of pesticide residues (organochlorine, organophosphate, and carbamate) were detected in 50 per cent of 1,500 water samples which were collected from many monitoring stations along the major rivers throughout the country, and in almost all of 1,300 samples of soil and sediment.

In the period 1989–1995, the Environmental Research and Training Center (ERTC) Thailand conducted a monitoring study on organochlorine pesticides in the Chao Phraya river (ERTC Thailand 1995). The water samples were collected from 21 stations, as shown in Figure 4.1.

The results of the study show that organochlorine pesticides such as BHC, aldrin, dieldrin, and DDT are present in river water. Relatively higher residue levels of aldrin were detected in the upstream sections. Figure 4.2 shows the concentration of organochlorine pesticides in the lower Chao Phraya river. The results seem to relate to the fact that the basin of the upstream Chao Phraya river is mainly an agricultural area. In some areas concentration of DDT is higher than 20 ng/l in total due to the usage of DDT for a malaria eradication programme until 1994.

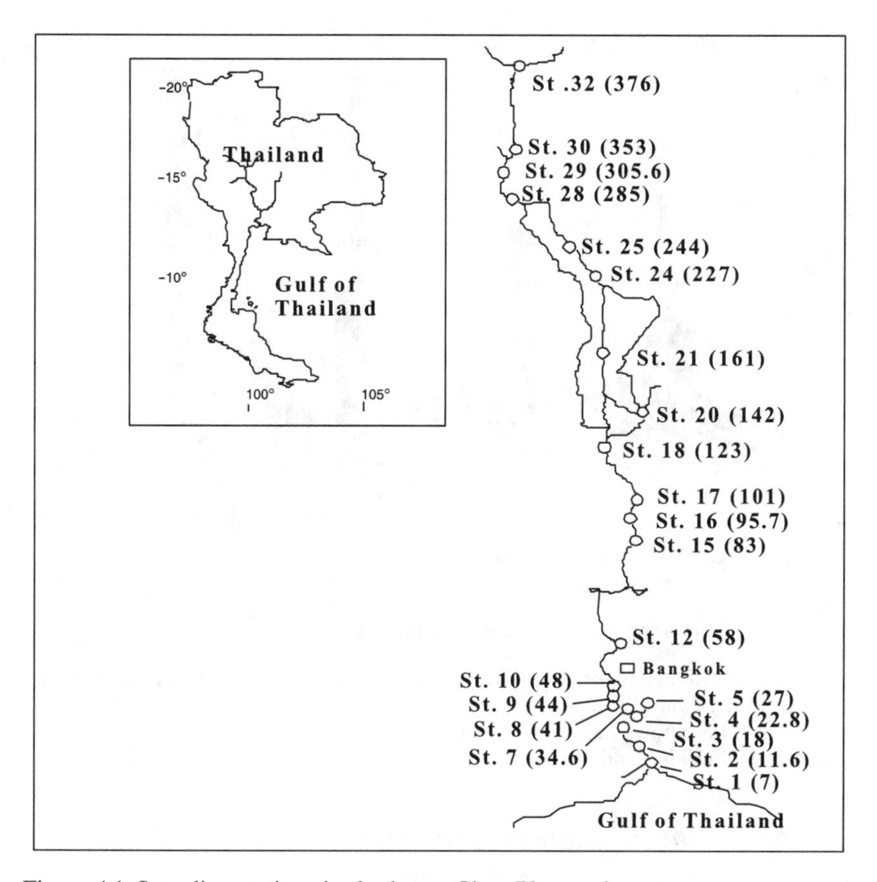

Figure 4.1 Sampling stations in the lower Chao Phraya river
Note: Figures in parentheses show the distance (km) from the river mouth.
Source: ERTC Thailand (1995)

However, the trend of concentration of organochlorine pesticide decreased in 1993. For example, the residual concentrations of BHC, heptachlor, dieldrin, and DDT are considerably lower than the Surface Water Quality Standard, as shown in Table 4.2.

The Thai government banned many organochlorine pesticide compounds for use in agriculture, as shown in Table 4.3, except DDT, which was used mainly for mosquito control in specific areas until 1994. To replace DDT, the Department of Communicable Disease Control, Ministry of Public Health, has used deltamethrin and trihalothrin since 1995. However, the concentration of DDT residues were lower than the maximum residue limit (MRL) for aquatic animals as recommended by the Ministry of Public Health of Thailand.

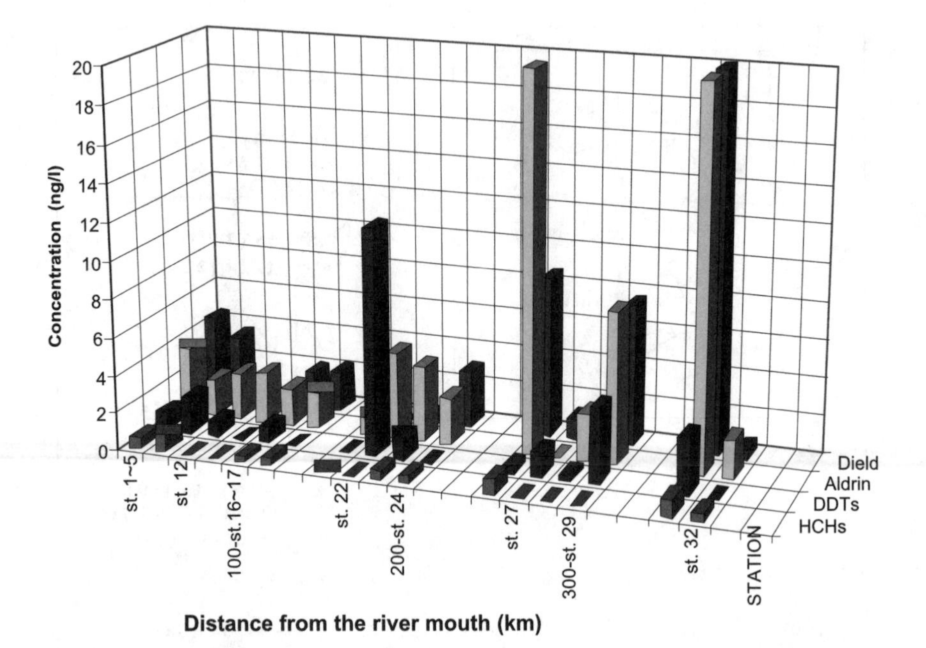

Figure 4.2 Spatial changes of residue level of total HCHs, total DDTs, aldrin, and dieldrin in the Chao Phraya river
Note: Each bar shows the median value.
Source: ERTC Thailand (1995)

Organochlorine pesticides in green mussels

In most cases, the ultimate environmental sink of agricultural pesticides is the coastal water. The "mussel watch" concept utilizes the concentrations of pesticides and other contaminants in mussels as an indicator of the level of stress that is being placed on the biological systems in coastal waters. Mussels have been shown to have many advantages as bioindicators for monitoring trace substances in coastal waters because of

Table 4.2 Comparison of concentration of pesticides found in the Chao Phraya and Bang Pakong rivers with Surface Water Quality Standard in Thailand

Type of pesticides	Standard (ng/l)	Chao Phraya (ng/l)	Bang Pakong (ng/l)
BHC	20	6.86	2.46
Heptachlor	200	21.02	0.47
Dieldrin	100	0.18	0.13
DDT	1,000	2.94	0.08

Source: ERTC Thailand (1995)

Table 4.3 Organochlorine pesticides banned and/or restricted by the Thai government

Chemical	Year effective
HCHs	1980
Endrin	1981
DDTs	1983
Aldrin	1983
Endrin	1983
Toxaphene	1983
Heptachlor	1988

Source: Department of Agriculture, Thailand (1988)

their wide geographical distribution, sessile lifestyle, easy sampling, tolerance of a considerable range of salinity, resistance to stress, and high accumulation of a wide range of chemicals. The green mussel (*perna viridis*) has a wide geographical distribution in the Asia Pacific region and is a commercially valuable seafood on a worldwide scale.

The concentrations of organochlorine pesticides in green mussels from the Gulf of Thailand in 1989–1996 were monitored by the ERTC Thailand – Figure 4.3 shows the sites for the mussel samples. Organochlorine pesticide compounds such as aldrin, dieldrin, HCH-isomer, and DDT-isomers were found in all stations during the period 1989–1990. Despite the fact that the use of DDT for agricultural purposes was banned in 1983, it was still used for malaria vector control by the Ministry of Public Health until 1994. Overall trends in relative concentrations of pesticides in mussels are shown in Table 4.4. These trends show the relative importance of each pesticide, while also indicating the impact of banning the use of certain pesticides on their environmental concentrations.

Hexachlorobenzene (HCB) is produced as a by-product material of several chlorinated hydrocarbons, such as in the production of tetrachloroethylene, trichloroethylene, chlorine, and vinyl chloride. Aldrin and dieldrin were formerly used as an insecticide against termites and soil insects; aldrin readily breaks down to dieldrin in living systems. The trend of aldrin and dieldrin concentrations in mussels from 1989 to the present has been a gradual decrease due to the ban on production of these compounds in 1983. During the 1993–1996 period, p,p′-DDE was the most frequently found compound among organochlorine pesticides. The major source of p,p′-DDE was metabolic transformation of p,p′-DDT to p,p′-DDE-isomer under oxidation conditions.

Occupational problems

Farmers use pesticides intensively and there is little control of the amounts used or the application frequency. Thongsakul (1990) states that

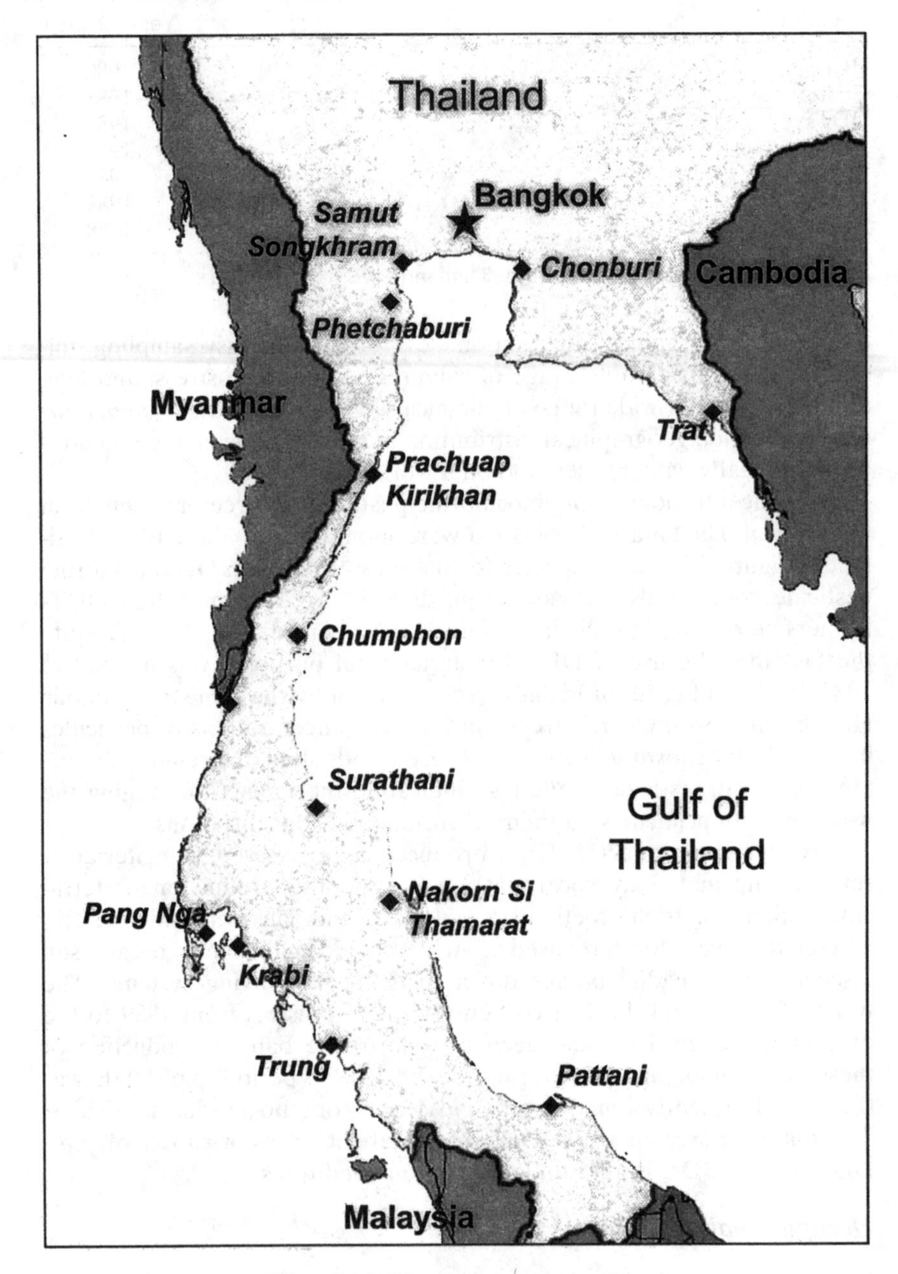

Figure 4.3 Sites for mussel samples

Table 4.4 Trend of organochlorine pesticide residues, 1989–1996

Year	Trend of organochlorine pesticides	References
1989	DDTs > aldrin > dieldrin > HCB > heptachlor > HCHs	Siriwong *et al.* 1991
1990	DDTs > dieldrin > aldrin > HCHs	Tabucanon *et al.* 1990
1991	DDTs > aldrin > dieldrin > Heptachlor	Ruangwises *et al.* 1994
1993–1995	β-HCH > α-HCH > δ-HCH > endosulfan II	Boonyatumanond *et al.* 1996
1996	Cis-chlordane and α-HCH	

the application of pesticide mixtures is a common practice. Pesticide mixtures often consist of several pesticides and are mixed without instruction or knowledge about the effectiveness or possible side-effects of the combinations. Spraying frequently is crop-dependent, but can reach high levels. The farmers' lack of awareness is seen as one major reason for pesticide problems. Several studies about farmers' awareness conducted in Thailand concluded that more than half the farmers applied dosages higher than recommended on the product labels. Almost all the farmers regularly mixed two or more pesticides for one application. Possible reasons can be to save time or the common belief that pesticide mixtures are more effective. Decisions about pesticide mixtures are made either by retailer recommendations or according to the common practice in the area (Songsakul 1991).

An assessment of health hazards related to pesticide use in agricultural production raises some difficulties. When poisoning cases occur, it is difficult to identify without doubt a specific pesticide as the source of poisoning. At the same time, many poisoning cases are never reported to the hospitals and will therefore never appear in the official occupational poisoning statistics. Generally speaking, most farmers lack real knowledge of the potential hazards that pesticides may cause for themselves and the consumer. Often, they spray pesticides frequently and harvest their crops for marketing before the end of the recommended waiting period; many farmers also do not pay sufficient attention to pesticide levels and protective clothing.

In Thailand, the Division of Epidemiology of the Ministry of Public Health has the primary responsibility for collecting data on poisoning. Pesticide poisoning is considered as one of the major occupational health hazards. During the 1988–1996 period the Ministry of Public Health reported 2,000–5,000 pesticide poisoning cases and 30–60 patients died per year. Because these data rely on case reports of governmental hospitals and some private clinics, the actual number of poisoning cases is assumed to be understated (Sinhaseni 1990).

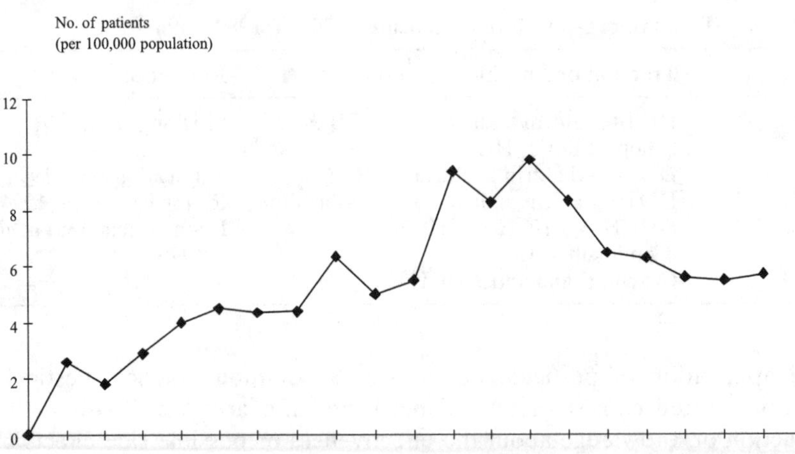

Figure 4.4 Pesticides poisoning cases, 1977–1995
Source: Ministry of Public Health, Thailand (1996)

The number of occupational poisoning cases officially listed has decreased in the recent past. The Ministry of Public Health reported that from 1977 to 1995 the rate of pesticide poisoning cases increased from 2.33 patients per 100,000 of the population in 1977 to 9.8 patients per 100,000 of the population in 1989, then decreased to 5.66 patients per 100,000 of the population in 1995, as shown in Figure 4.4. There are no apparent reasons for this reduction because the amount of pesticides imported and used has increased and no radical changes in the type of pesticides used or the application technology chosen have taken place.

Food safety problems

Pesticide residues in food for human consumption and export have been a great concern in Thailand. Misuse and mishandling of pesticides by farmers and users could cause human health hazards. The pesticide residue problem is more serious in the production of vegetables and fruits, and recently in other food. The Food and Agricultural Organization (FAO) and the World Health Organization (WHO) have jointly developed standards for pesticide residues in rice based on protection of human health – these are shown in Table 4.5.

Pesticide residues in rice were monitored by the Department of Agriculture during the period 1995–1996; 171 samples of rice were collected from different regions in provinces in Thailand, as presented in Table 4.6. Organochlorine pesticides – such as BHC, aldrin, dieldrin, heptachlor,

Table 4.5 International standards for MRL and ADI of pesticide residues in rice

Pesticide	MRL (mg/kg)	ADI (mg/kg body weight)
Lindane	0.5	0.008
Heptachlor and heptachlor epoxide	0.02	0.0005
Aldrin and dieldrin	0.02	0.0001
Total DDT	0.1	0.02
Endrin	0.02	0.0002
Chloropyrifos-methyl	0.1	0.1
Malathion	8.0	0.02
Dichlorvos	2.0	0.004

Source: FAO/WHO (1992)

epoxide, endrin, DDT and its metabolites, and endosulfan – were found in 132 out of 171 samples or 77 per cent of analysed samples. These pesticides were found in all the regions of country sampled, showing a widespread problem with organochlorine pesticides. Organophopshate pesticides – such as monocrotophos, diazinon, and malathion – were mostly found in the central, eastern, and southern regions. Carbamates were found primarily in the central and north-eastern regions.

The study also conducted a comparison of pesticide residues accumulated in Thai rice and Japanese rice (see Table 4.7). The rice samples were collected from the experimental stations of the Ministry of Agriculture and cooperatives and private sector sites at Chiang Mai, Chiang Rai, Lumpun, and Bangkok, and from imported Japanese rice. When evaluated, organochlorine pesticides were found in the rice samples from both the private sector and rice grown in the research institutes. In comparison, the imported Japanese rice was found to have no pesticide residues. Additionally, the rice planted by the private sector had higher pesticide concentrations than other samples. However, the concentration of pesticide residues in the rice was found to be within the MRLs recommended by the FAO/WHO (1992).

Legal and institutional framework

Legal framework

Factories Act (1995)

The Factories Act empowers the Ministry of Industry (MOI) to control the establishment and operation of factories. The MOI can issue regulations imposing limits on discharges of air pollutants, effluents, or wastes

Table 4.6 Sampling sites, number of rice samples, and concentration of pesticide residues found in different regions of Thailand

			Pesticide residues found					
			Organochlorine pesticides		Organophosphate pesticides		Carbamate	
Sampling sites	Number of provinces	Total number of samples	Number of samples	%	Number of samples	%	Number of samples	%
Central region	9	47	28	59.6	1	2.1	3	6.4
Eastern region	5	23	21	91.3	4	17.4	–	–
North-eastern region	9	45	40	88.9	–	–	7	15.6
Northern region	5	25	20	60.0	–	–	–	–
Southern region	6	31	23	74.2	2	6.5	–	–
Total	34	171	132	77.0	7	4.1	10	5.8

Source: Department of Agriculture, Thailand (1996)

Table 4.7 Comparison of organochlorine pesticides found in rice planted by the private sector, research institutes, and imported from Japan

Source	Number of samples	Organochlorine pesticides found	
		Number	%
Rice planted by private sector	13	11	84.6
Rice planted by research institutes	30	12	40.0
Imported rice from Japan	2	–	–

Source: Department of Agriculture, Thailand (1995)

from pesticide factories. The MOI has also established the Office of Hazardous Substances, assigned to work solely on toxic and hazardous materials.

Hazardous Substances Act (1992)

The creation of the Hazardous Substances Act can be regarded as a major change in crop protection policy in Thailand – in 1992 it replaced the Poisonous Articles Act of 1967. It is the main law that regulates pesticide use in agriculture, in addition to about 20 existing laws related to the control of chemicals in all areas of pesticide use (Boon-Long 1995). Under the umbrella of this Act, tax policy, import, trade, and use regulations for pesticides have been formulated. According to the Act, for the regulation of production, import, export, and possession, hazardous substances are categorized into four groups:

- those which are in compliance with the principles and procedures of the Act
- those which must be reported to the authorities
- those which require a permit
- that which are legally banned.

No registration is necessary for chemicals in the first category, while those in the second and third categories need to be registered before manufacture and/or import. In this Act, the Hazardous Substances Committee has been enlarged, and the Hazardous Substances Board represents the legislative arm of the Act. There are several subcommittees working on more specific issues of hazardous chemicals. One of the new aspects enclosed in the Hazardous Substances Act is the responsibility given to all persons being in possession of a hazardous chemical for damages to persons, animals, crops, and the environment.

Since 1991, the Department of Agriculture has declared the phased registration system according to the code of conduct to be the new system for pesticide registration. However, the ministerial decree for enforcing phased registration came into action in 1995, indicating a somewhat slow process of implementation. All types of pesticides are controlled by the Act. The regulatory process involves three steps: obtaining the registration licence; obtaining an import, manufacturer, and/or retailer licence; and market inspection.

According to the phased registration system, there are three groups of pesticides for the registration process:

- pesticides which have never been registered in any country must be registered in phase 1
- pesticides already registered in other countries but not in Thailand may be registered in phases 1 or 2
- pesticides which are registered in other countries and in Thailand may be accepted in phases 2 or 3.

Each pesticide has to be tested in Thailand for risk-benefit assessment and effects on humans and the environment. If a product has already been tested elsewhere, only missing toxicological or bio-efficacy data are requested. After testing, the Subcommittee for Registration decides on the registration licence, which has to be issued for every formulation. Once issued, licences may be revoked if evidence of hazards can be proved, but this rarely occurs. It is unclear how licences for products already registered can be issued for three years. Currently the licences are only issued on a yearly basis, thus giving the regulatory division some chance to deny the import if evidence occurs according to the registration requirements. The fees for permits were increased in the Hazardous Substances Act, although the actual fees (notified in the Hazardous Substances Decree of November 1994) are much lower. Currently, the cost for pesticide registration is 1,000 baht (the upper limit is fixed at 5,000 baht according to the Act); additional fees for import and trade licences exist.

In order to ban a certain pesticide the subcommittee is responsible in the first stage and has to prove the reasons for a ban. Then the Hazardous Substances Board makes the final decision for banning the active ingredients.

Tax policy

In general, the total import taxes consist of import duty, business tax, and municipal tax and are based on CIF-price value. The tax structure related to pesticides has been favourable compared to other inputs and has therefore helped to keep pesticide prices low. The import duty on pesticides does not take into account the hazardousness of a pesticide. Before 1991, effective total tax rates for pesticides were 6.9 per cent, compared to 32.4 per cent for fertilizer and 27.6 per cent for agricultural machinery (Waibel 1990).

In cases of pesticides which are identified as being for agricultural use only, they have been exempted from import duty and business and municipal taxes since 1991 (Customs Department 1995). Taxes may occur for some ingredients in pesticide formulation which can be used for other than agricultural purposes. Starting from 1995, import duty for fertilizer has also been favoured, with a reduction of import duty to 10 per cent (formerly at 30 per cent). This tax exemption can be clearly defined as an indirect subsidy for pesticide imports and pesticide prices. It can also be seen as a subsidy for hazardous products which are cheap on the world market and do not face taxation according to their hazardousness when imported to Thailand.

Import, trade, and use regulations

The Regulatory Division of the Department of Agriculture (DOA) is in charge of the process of pesticide registration and the supervision of laws controlling the use of pesticides. At the same time, the responsibility for controlling the pesticide market in terms of quality control and residue analysis, necessary for the registration process, rests on the Toxic Substances Division of the DOA. There are two designated national authorities for the PIC (prior informed consent) scheme. The Department of Agricultural is in charge of all chemicals used in agricultural production, and the Pollution Control Department is responsible for all other chemicals.

At present, all pesticides require registration prior to import, manufacturing, and distribution. Currently there are 298 active ingredients registered in Thailand, which add up to a considerably higher number of product names (2,258 names in 1991), and the trend is still rising. Grandstaff (1992) states that there also exist illegal repackers who never applied for a permit and who are therefore not under regulatory control.

From the viewpoint of the implementing agency the registration process raises some concerns. On the one hand there exists a strong dependency on other divisions of the DOA for registration-related research, and requirements which are not classified as priority issues are therefore limited to a very small sample size and only a few quality control tests can be conducted. This is a shortcoming of the current legislation, as the quality of pesticides is a major concern *vis-à-vis* the inappropriate use of pesticides.

Starting in July 1995, retailer training has been made compulsory for retailers to retain their licences. Retailers are required to attend a training course within 18 months. Prior to July 1995, retailer training existed but it was not compulsory and it was partly initiated by pesticide companies. Training lectures are arranged by the DOA as well as the Thai Crop Protection Association in cooperation with the International Group of

National Associations of Manufacturers of Agrochemical Products (GIFAP). The training focuses mainly on the introduction of the new Act and also includes aspects of entomology, plant pathology issues, storage regulations, and safe use. However, the success and the sufficiency of these two-day training modules may be questioned. The impact of the retailer on farmers' pest management decisions is of high relevance, as retailers are often the only or main source of pesticide recommendations and information for the farmer (Khuankaew 1995). The lack of market transparency can be regarded as a pesticide supporting factor.

Institutional framework

The Agricultural Department

Two departments within the Ministry of Agriculture and Cooperation (MOAC) cover almost all aspects of agricultural crop production policy. They share more than 50 per cent of the MOAC's budget for research and extension.

The DOA is in charge of all agricultural research projects and responsible for developing technologies, which are supposed to be tested and transferred to the farmers by the Department of Agricultural Extension (DOAE). The DOAE is in charge of extension work and the formulation of strategies for technology dissemination, according to national policy targets. Several divisions within each department focus separately on various aspects of agriculture. The government research allocation is focused more on importable commodities.

Plant protection issues are dealt with in the Plant Protection Service Division (PPSD) of the DOAE and the recently formed Bio-control Center, as well as several divisions of the DOA (regulatory, toxic substances, entomology, etc.). Integrated pest management (IPM) related research or extension work falls under no special division but instead is part of the overall agricultural policies. In the DOA IPM research projects for different crops were conducted over several years (the rice project started in 1983, cotton in 1981, sugarcane in 1985, and fruits in 1989). Apart from these projects most research conducted by the DOA in the field of pesticides concerns pesticide efficacy and application techniques. From 1981 to 1988, a project on surveillance and early-warning systems was carried out nationwide. In 1980 a bio-control subdivision was established in the DOA as well as in the DOAE. The DOAE has offices at the regional, provincial, district, and subdistrict levels.

The Environmental Department

There are three departments within the Ministry of Science, Technology, and Environment (MOSTE).

- The Office of Environmental Policy and Plan (OEPP) is in charge of the strategic plan for natural resource management. This management includes the environmental quality of soil, water, and sediments in different classifications of watershed.
- The Pollution Control Department (PCD) focuses on the development of environmental quality standards that include setting limits on the acceptable levels of pesticides in surface water and groundwater.
- The Department of Environmental Quality Promotion (DEQP) is in charge of information exchange, the public participation programme, research and development, and technology transfer. The ERTC is designated for monitoring activities, methodology development, and training related to pesticide residues. Different programmes are conducted by the ERTC for the monitoring of pesticides, such as the mussel watch programme, monitoring of organochlorine pesticides and polychlorinated biphenyls in the Chao Phraya and Mae Khong rivers, and monitoring volatile organic compounds in groundwater.

Crop protection policy

Integrated pest management (IPM) has been recognized as a high priority in crop protection policy in Thailand. Increasing attention has been given to IPM methods in recent years, but the concept of IPM in terms of using several management and pest control possibilities at the same time has not yet been sufficiently applied in the field. IPM could be a solution to more sustainable agriculture. Major constraints are present for the implementation of IPM or anything other than chemical crop management practices. Most importantly, there is a lack of sophisticated and adaptable technology for most crops to achieve a more widespread adoption of IPM. There is also a lack of research on IPM-based production systems as well as poor development of extension and training tools for the transmission of knowledge on IPM methods at farmers' and extension levels. Consequently, there is a lack of interest and knowledge at the farmer level in IPM methodologies.

A ranking of the crop protection policies shows a clear priority on pesticide-related issues like safe use, outbreak budget, and regulatory policies. Education and training curricula along with IPM training and school education are highlighted as having some influence on discouraging pesticide use. However, stronger incentives for using pesticides as the sole pest management tool still persist in the farming community. For implementation of IPM activities more research is needed on the design of adaptable and successful IPM systems for various crops. Only when such adaptable IPM systems are developed can successful implementation be expected.

Information management

Information is essential to rational decision-making about the use of pesticides. There are several kinds of information needed at the policy level, at the research level, and at the farmer level. Two kinds of information related to pesticide use are essential: the benefit and cost of the use of a certain pesticide; and general information about possible alternatives to its use. The main information about pesticide use available from the DOA includes efficacy tests and assessment of crop losses. However, the pesticide recommendations of the DOA seem to be far too complex for daily use and are not a major source of information for farmers' decision-making in crop protection. These measures encourage pesticide use, as the available information promotes the use of pesticides. Waibel (1990) states that there is a lack of information on the danger of application and handling of pesticides as well as about the quality and formulation of pesticides, the production date, and the ingredients of pesticides.

Information about pesticide quality can seriously affect pesticide use levels. Most farmers apply increased dosage due to their experience that the recommended amount was found to be in effective. It can be said that a lack of information regarding the actual amount of active ingredient applied does not support good management decisions and pest control. Information for improving the quality of agricultural products in the form of an internet-based website was established by the DOA in 1998. The main objectives of this website are:

- to formulate an information system for distributing research results and information which will be of benefit to researchers, exporters, farmers, etc.
- to act as a medium for providing information about the DOA to the public
- to act as the focal point in exchanging views, opinions, and news related to agriculture through the DOA home-page, internet, bulletin board system (BBS), and hotline.

Public participation

The major goals of public participation in pesticide and environment programmes in Thailand are to ensure an open and accountable planning and programme process, to provide information to the wider community, and to create public understanding of a proposed action, programme, or plan related to the usage of pesticides with two-way communication to promote understanding and problem-solving. In this respect, various training programmes are conducted by the MOAC, the MOSTE, and some NGOs.

On the provincial level, several NGOs conduct training and educational work in the field of alternative management systems. These NGOs are most active in the areas of rural community development, especially integrated rural development. Activities such as the setting up of rice banks, fertilizer banks, and pesticide banks were common in the early 1980s. This rural development work expanded into small-scale water resource development, including mixed farming. With a growing interest in environmental issues as well as in natural resources, an increasing number of NGOs began promoting the concept of sustainable agricultural development. Environmental NGOs include the Society for the Conservation of National Treasures and Environment (SCONTE), the Siam Society, the Project on Ecological Recovery (PER), and the Thai Environment and Community Development Association (TECDA). The Life-Long Education Foundation has been active in producing television programmes which promote environmental education. Most of the programmes of NGOs are focused on farmers or farmer groups, with some interested in IPM education in schools.

Consumer awareness of possible effects related to pesticides creates a growing market for controlled or pesticide-free produced crops. NGOs working in the field of pesticide-free or organic food production, such as the Royal Project and Mae Fah Luang Foundation, are designing rules and regulations for the production of these crops, thus creating the necessary foundation for consumer acceptance. However, improvements in the current design can also contribute substantially to limit the ongoing support for pesticides in Thailand.

The problems related to the use and ongoing support for the use of pesticides in rice as the main crop protection strategy have been recognized by different sectors in Thailand. The promotion of agricultural exports stimulates overall pesticide use that emphasizes the quality and appearance of the crop and export standards. A stronger role for agricultural producer interest groups in the registration process would be a desirable step towards more unbiased decision-making. As shortcomings in the current implementation of pesticide legislation have been identified as pesticide supportive, a critical assessment of forces and structures within the governmental system could be a good step. Higher regard for the incorporation of economic instruments into the crop protection policy is essential in order to limit pesticide use effectively to the social optimum.

REFERENCES

Boon-Long, J. 1995. "Chemical legislation and infrastructure in Thailand", paper presented at the Conference on International Trade of Dangerous Chemicals, Brussels, 5–7 July.

Customs Department. 1995. *Customs Tariffs*. Bangkok: Ministry of Industry.

Grandstaff, S. 1992. *Pesticide Policy in Thailand*. Bangkok: Thailand Development Research Institute.

Mills, L. 1994. *Consultant Report on Pesticide Use in Various Countries*, London.

Sinhaseni, P. 1990. *Regional Pesticide Review*. Bangkok: Thailand International Development Research Center, Chulalongkorn University.

Songsakul, T. 1991. "Effects of urbanization and environment on vegetable cultivation: A case study of Village No. 3, Ban Mai subdistrict, Pathumthani province", masters thesis, Asian Institute of Technology, Bangkok.

Tayaputch, N. 1988. "Pesticide residues in Thailand", in Teng, P. and K. L. Heong (eds) *Pesticide Management and Integrated Pest Management in Southeast Asia*, USAID.

Tayaputch, N. 1992. "Fate of pesticides in tropical environment of Thailand", PhD thesis, University of Agriculture Tokyo, Tokyo.

Thongsakul, S., D. Wechakit, J. Ek-Amnuay, and P. Ratanasien. 1990. *A Survey of Tangerine Growers Concerning Application Spray Equipment and Their Problems*. Bangkok: Pesticide Application Research Section, Entomology and Zoology Division, Department of Agriculture.

Waibel, H. 1990. "Pesticide use and pesticide policy in Thailand", paper presented at the Workshop on Environmental and Health Impacts of Pesticide Use in Rice Culture, Los Banos, Philippines, 28–30 March.

Wood Mackenzie & Co. 1993. *East Asian Agrochemical Markets*, consultancy reports. London: Wood Mackenzie & Co.

Case studies of water resources

5

Governance scenario for water resources in Malaysia

Abdul Rashid Ahmad and Hasnah Ali

Challenges in managing water resources

The specific problems related to fresh water in Malaysia are numerous – these pertain to both the quantity of water available and its quality. As shown in Table 5.1, the increasing water demands parallel the population growth, raising concerns about the quantity of clean fresh water available. It is interesting to note that usable water resources are abundant in Malaysia, with an average annual precipitation of 2,985 mm. The country has adequate drinking water resources for the estimated population of 23 million in the year 2000. However, industrial development has contributed significantly to the problems of pollution by toxic and hazardous wastes, which pollute surface waters in rivers and lakes. These wastes, which include acids, asbestos, and heavy metal sludge, are outputs of the metal-finishing industries, textile industries, gas processing, foundries and metalworks, and asbestos factories. Even with the abundant rainfall, preventive or conservation measures and remedial action are needed to ensure an adequate supply of clean water for the future.

Available water resources

In most states of peninsular Malaysia, Sarawak, and Sabah, surface waters are the main sources of drinking water, i.e. from rivers and reservoirs. The average estimated run-off of the whole country is 566 billion

Table 5.1 Projected domestic and industrial water demands (million m^3/year)

State	1980	1985	1990	2000	Population in 2000 (millions)
Perlis	7	9	16	37	1,835.9
Kedah	49	82	113	260	
Pulau Pinang	124	169	236	343	1,259.4
Perak	145	216	327	596	2,130.0
Selangor	470	658	787	1,201	4,711.7
North Sembilan	62	102	131	191	849.8
Melaka	30	43	61	112	598.9
Johor	159	258	338	578	2,731.5
Pahang	49	116	193	455	1,319.1
Trengganu	31	53	82	222	1,064.1
Kelantan	34	60	99	311	1,561.5
Peninsular Malaysia	1,160	1,766	2,383	4,312	18,061.9
Sabah	58	82	103	259	3,136.8
Sarawak	59	92	124	273	2,064.9
Malaysia	1,277	1,940	2,610	4,844	**23,263.6**
Raw water to Singapore	198	250	316	414	
Total	1,475	2,190	2,926	5,258	

Source: JICA (1982); Malaysia (1996)

cubic metres per year, and the average annual precipitation per capita is 50,000 cubic metres. There are 150 river systems with approximately 1,800 rivers and a total length of over 38,000 km. The main rivers and lakes of peninsular and east Malaysia are shown in Figure 5.1. These rivers are administered by the state governments. With the occurrence of unusually long dry spells, localized shortage of water may happen. Generally, in areas of high economic activity and population growth, it is likely that demand for water will exceed supply. Groundwater also comprises a major component of available water resources. In the state of Kelantan, located in the north-east of peninsular Malaysia, groundwater of the Kelantan plain is one of the sources of water. The groundwater resources of the plain are the most extensive in Malaysia, and the source of drinking water for the majority of the rural population.

Water quality issues

The availability of water resources is only half the picture – equal emphasis has to be laid on water quality. The rapid development of industries in Malaysia has contributed considerable pollution to the rivers of the country. Surface water pollution by organic matter is one of the

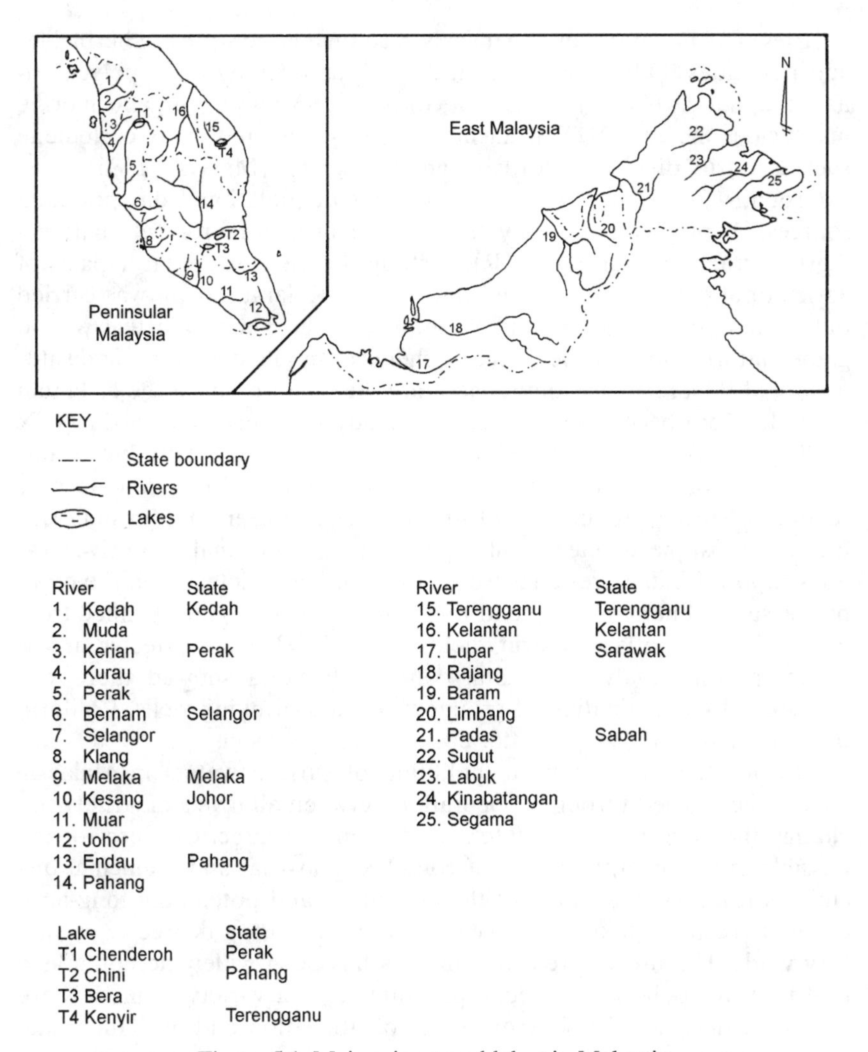

Figure 5.1 Major rivers and lakes in Malaysia

major problems in most areas of the country. Organic matter sources include industrial discharge (agro-based and non-agro-based), domestic discharge, domestic sewage (partially treated and untreated), and animal waste (mainly from pig rearing). Suspended solids, largely resulting from development activities (housing and urban) and deforestation due to logging, are another significant contributor to surface water pollution. Water pollution is particularly serious in areas of high population density and where there are significant numbers of agro-based and other industries. The major organic chemicals used on a large scale are herbicides and pesticides. Other toxic chemicals such as phenolics and polychlorinated biphenyls (PCBs) are also of concern. In Malaysia, organochlorine insecticides include DDT, hexachlorocylohexance, and many cyclodiene such as aldrin, dieldrin, chlordane, and heptachlor (Malaysia 1986).

Groundwater is not immune to quality degradation from pollution sources. A project funded by the Australian Centre for International Agricultural Research (ACIAR) to study the environmental impacts of agricultural practices on groundwater of the Kelantan plain was carried out (Aminuddin, Sharma, and Willet 1996); the main objective was to gather information on the effect of chemical inputs on the groundwater quality of the area. Rice and tobacco are the major crops of the Kelantan plain. In the tobacco-growing areas of sandy soils, high levels of NO_3-N in the groundwater were found to exceed the standard for drinking water quality. There was evidence the NH_4-N and NO_3-N concentrations were related to surface application of nitrogenous fertilizers. Endosulfan was found in most piezometer, drinking wells, irrigation canals, and river water samples. Lindane residues were also commonly detected in the cropping systems, but at low concentration. Some pesticides now banned from use, such as carbofuran, aldrin, dieldrin, and DDT, were detected at low concentration. Analyses of total coliform bacteria showed there was widespread contamination of groundwater and drinking wells. Coliform bacteria increased abruptly after each monsoon season.

The problems pertaining to pollution of surface waters in Malaysia have to be tackled through cooperation between all levels of society, including the government, polluters, and people at large. Concern for the possible environmental effects of complex industrial and municipal discharges requires that more subtle techniques and potentially long-term effects have to be taken into account to an increasing degree in regulatory work. The use of predictive models has been widely accepted as a tool in water pollution remedial programmes – a variety of models are available, designed for different types of utilization and available data. Less complex models are more often used in pollution control programmes. More scientific efforts are being initiated to gain further knowledge into load-response relationships for various water bodies. These relationships would then become widely available managerial tools for establishing pollution reduction strategies for many aquatic ecosystems.

Rational decision-making also requires sufficient and reliable data on both quality and quantity of water resources. The monitoring of water quality is carried out by several government departments and ministries: the Department of Environment (DOE), Department of Irrigation and Drainage (DID), Public Works Department (PWD), and Ministry of Health (MOH). In addition, various other institutions, both private and governmental, also undertake monitoring but on a limited scale and for specific purposes mainly relating to particular project activities. The National Electricity Board undertakes rainfall, river discharge, and sediment monitoring, mainly in areas where hydroelectric power generation projects are undertaken. The Geological Survey Department conducts various studies to evaluate groundwater potential in various parts of the country.

The Department of Environment currently operates 350 water quality monitoring stations in the 49 water quality control regions throughout peninsular Malaysia. A wide range of quality parameters are observed, including heavy metals and inorganics, radioactivity, physical and microbial parameters, and various organic pollutants such as herbicides and pesticides. Monitoring of groundwater mainly from boreholes and open wells is carried out regularly at 44 stations, mainly in the north-east and north-west of peninsular Malaysia. There are 94 river basin control regions in Malaysia: 49 in peninsular Malaysia, 21 in Sarawak, and 24 in Sabah. The DID maintains 69 principal and secondary water quality monitoring stations in peninsular Malaysia, 191 stream flow stations, 271 automatic and 752 manual rainfall recording stations, and 95 suspended sediment sampling stations throughout the country. The PWD and Ministry of Health jointly carry out monitoring of water supply and sanitation. In addition, the MOH monitors the quality of water in the distribution network. It also undertakes the monitoring of water supplies from ground and surface waters, particularly in areas where waterborne diseases are of concern. In summary, a wide range of data pertaining to quality and quantity of water are collected at various levels within the country.

National legislative framework in Malaysia

Upholding water quality and quantity are closely associated with environmental protection. An early form of legislative response to impending environmental problems in Malaysia was the introduction of the Water Enactment in 1920. The Water Enactment was followed by other legislation which had relevance, though often indirect, to the environment. As development in the country progressed rapidly, the mainstream water-related laws appeared to be directed at environmental protection. The standards for water quality are established by the DOE. These standards were adopted by various government departments for water quality controls. The MOH is responsible for advising on standards for monitoring of water-related microbiological contamination with respect to health problems. The World Health Organization (WHO) standards were considered during the establishment of national standards; in most cases, the national standards are directly adopted from WHO standards.

Major state laws corresponding with state matters involving water are the Water Enactment, the Rivers and Drainage Enactment of Kelantan, the River Obstruction Enactment of Johore, the Water Supply Enactment, and the Water Authority Enactment of Penang and Malacca. Water resources development and management programmes in the states of peninsular Malaysia are executed in accordance with numerous laws. There are several areas, directly or indirectly associated with water quality and quantity, where water-related laws are involved.

River and water management

The Water Enactment is the fundamental law with respect to river and water management. It contains provisions for property in rivers, prohibition of diversion of water from rivers except under licence, and restrictions on construction of walls and building on the banks of rivers or within flood channels, as shown in Table 5.2.

Water supply

The Water Supply Enactment prevails in states other than Penang and Malacca. It deals with declaration of water supply areas, water supply services, and imposition and determination methods of water rates. The Penang Water Authority Enactment provides responsibility for the Penang Water Authority to manage water supply throughout the state; similarly, the Malacca Water Authority manages water supply in accordance with the Malacca Water Authority Enactment, as shown in Table 5.3.

Water pollution control

The legislation most related to water quality and quantity is the water pollution control regulatory framework (WPC). The WPC is to be implemented under several laws, such as the Environmental Quality Act, the Water Enactment, the Local Government Act, and the Street, Drainage, and Building Act. The Water Enactments of most states, except Perlis, Kedah, and Pahang, were amended to include clauses providing for the prohibition of water pollution. According to this amendment, no person shall discharge polluting matters into a river except in accordance with the terms and conditions of a licence issued by the state secretary. The Environmental Quality Act (1974) is the basic law providing for pollution control and environmental improvement; it also places restrictions on pollution of inland waters. Water pollution control

Table 5.2 Legal provisions for river and water management

Law	Purpose of law	Involved provisions
Water Enactment	Control of rivers and streams	– Property in rivers – Prohibition of acts affecting rivers except under licence – Prohibition of diversion of river water except under licence – Prohibition of pollution of rivers – Restriction on construction of walls and buildings on banks of rivers or within flood channels
Mining Enactment	Mining	– Control of and property in water – Control of interference with river banks – Licence for use of water – Purification of water
Geological Survey Act	To regulate and control geological surveys, to establish geological archives	– Development of certain wells to be notified – Certain excavations to be notified
Rivers Obstruction Enactment	Protection of rivers, streams, and estuaries against damage or obstruction by fallen timber	– Punishment for felling timber into river – Collector may order riparian owner to remove timber

Source: Environmental Quality Act, 1974; Merchant Shipping Ordinance 1952; F. M. Ord. 70/1952; Sabah Laws Forest Enactment 2/1968

in a local authority area is vested under the Local Government Act 1976, as shown in Table 5.4.

Watershed management

The control of silt and erosion and conservation of hill land are carried out under the provisions of the Land Conservation Act. The Forest Enactment provides for the procedures for constituting a reserved forest. The National Park Act provides procedures for the establishment of national parks, as shown in Table 5.5.

Hazardous waste management

Currently there are three specific regulations gazetted under the EQA 1974 to control toxic and hazardous wastes:

Table 5.3 Outline of legal provisions for water supply

Law	Purpose of law	Involved provisions
Water Supply Enactment*	Water supply	– Notification of a water supply area – Imposition of water rate – Power to make rules as to manner of determining the water rate – Private water supplies – Meaning of domestic supplies – Additional rate where private service is provided – Exemption of water rate in certain cases – Trade supplies – Charges for trade and private supplies – Inside services to be furnished before water laid on – Private services to be laid in accordance with rules – Supply and control of meters – Application to water supply areas outside town board areas
Penang Water Authority Enactment (Pulan Pinang)	To establish a corporate body by the name of Penang Water Authority to manage the water supply within the state of Penang	– The membership of the authority – Functions and duties of the authority – Transfer to the authority of the government water undertakings – Transfer to the authority of the City Council of George Town water undertakings – Establishment of the fund – Balancing of revenue accounts – Accounts and audit – Power to borrow – Power to make loans – Compulsory acquisition of land – Power to prescribe sources of water – Private water supplies – Trade supplies – Inside services to be furnished before water is laid on
Malacca Water Authority Enactment (Melaka)	Establishment of the Malacca Water Authority and matters connected therewith	– Establishment and incorporation of Malacca Water Authority – Constitution of the authority – Terms and conditions of appointment of members – Functions and duties of the authority – Funds and revenues of the authority – Transfer and acquisition of properties

Source: Federal Laws Environmental Quality Act (EQA) 1974, Malaysia Act 127; Sabah Laws Town and Country Planning Ordinance Cap 141 Vol. IV
* Applicable to states other than Pulan Pinang and Melaka.

Table 5.4 Outline of legal provisions for water pollution control

Law	Purpose of law	Involved provisions
Environmental Quality Act	Prevention, abatement, and control of pollution, enhancement of the environment, and purposes connected therewith	– Director-General and other officers – Establishment of the Environmental Quality Council – Licences – Prescribed premises to be licensed – Prohibition against causing premises to become prescribed premises – Requirement and approval of plans – Power to specify conditions of discharge – Restriction on pollution of inland waters – Power to prohibit use of any material or equipment – Power to require occupier to install, operate, repair, etc. – Power to prohibit or control licensed persons from discharging of wastes in certain circumstances – Delegation – Power to make regulations
Water Enactment*	Control of rivers and streams	– Prohibition of pollution of rivers
Local Government Act	Ensuring uniformity of law and policy to make a law with respect to local government	– Committing nuisance in streams, etc. – Pollution of streams with trade refuse, etc. – Local authority may recover for work done – Nuisances to be abated
Street, Drainage, and Building Act	Street, drainage, and building in local authority areas in west Malaysia, and for purposes connected therewith	– Local authority to make public sewers – Local authority to construct and maintain drains and watercourses – Not to cause nuisances

Source: Federal Laws Merchant Shipping Ordinance 1952; F. M. Ord. 70/1952; Malaysia (1991); Malaysia (1996); Sabah Laws Local Government Ordinance 11/1961

* Additional Clanse 7(A) of the Waters Enactment, excluding Perlis, Kedah, Pahang, and Kelantan. In Kelatan a similar provision is found in section 8A of the Rivers and Drainage Enactment.

Table 5.5 Outline of legal provisions for watershed management

Law	Purpose of law	Involved provisions
Forest Enactment	Forests and forest produce	– Constitution of reserved forest – Proclamation by district officer – New buildings and cultivation prohibited after proclamation – Inquiry by district officer – Regulation of privileges – Acquisition of alienated land for inclusion in a reserved forest – Notification declaring reserved forest – Power to stop way or watercourse in a reserved forest – Acts prohibited in a reserved forest – Offences in a reserved forest – Power to compound forest offences – Charge on forest produce for money due to government
Land Conservation Act	Conservation of hill land and protection of soil from erosion and the inroad of silt	– Declaration of hill land – Prohibition of short-term crops except under permit – Restrictions on clearing and cultivation of hill land – Control of silt and erosion – Maintenance of works – Power to make orders and nature of orders – Power to cause effect to be given to orders and recovery of cost – Liability of owner or occupier for acts done on land
Protection of Wildlife Act	Protection of wildlife and for purposes connected therewith	– Declaration of wildlife reserves and sanctuaries – Permits to enter wildlife reserves and sanctuaries – Prohibition of certain acts in wildlife sanctuaries
National Parks Act	Establishment and control of national parks and matters connected therewith	– Establishment of national park – National Parks Advisory Council – National park committees – Occupation of land within a national park – General prohibition of mining within a national park

Source: Sabah Laws Mining Ordinance 20/1960; Federal Laws Geological Survey Act (GSA) 1974, Malaysia Act 129

- the Environmental Quality (Scheduled Wastes) Regulation 1989 P.U.(A)139
- the Environmental Quality (Prescribed Premises – Scheduled Wastes Treatment and Disposal Facilities) Regulation 1989 P.U.(A)140
- the Environmental Quality (Prescribed Premises – Scheduled Wastes Treatment and Disposal Facilities) Regulation 1989 P.U.(A)141.

The waste producer is obliged to pack, classify, and label the waste before delivery to the transporter, accompanied by a consignment note in the form prescribed by the EQR (Scheduled Wastes) 1989. The transporter is also required to complete another form prescribed by the same regulation, to obtain an acknowledgement of delivery to the treatment facility, and to return the duly acknowledged form to the waste producer.

Sewage and industrial effluent

The discharge of effluent, both on to land and into inland waters, needs to be regulated in order to protect the environment. Under the EQR (Sewage and Industrial Effluents) 1979 there is a total prohibition on the discharge of any effluent waste and sludge in or on the surface of any land without the prior written permission of the Director-General of the DOE. Certain substances, such as inflammable solvent, tar, sawdust, timber, and human and animal waste, are completely prohibited to be discharged into inland water bodies. Under the same regulation, discharge of effluent into inland waters is allowed if it is within an allowable quantity per day with substances/contaminants within allowable limits.

The governance framework in Malaysia

The governance framework within Malaysia comprises two main components. Firstly, the national legislative framework provides a complex operating mechanism for day-to-day operational purposes. Since Malaysia is a federal state, the laws at the national level are complemented by a system of state laws and enactment. Secondly, on a broader scale a range of international treaties and institutions provide the backdrop for designing policies at a national level.

National framework

The Malaysian government realized the growing environmental problems in the country, passed the Environmental Quality Act, and established the Department of Environment under the Ministry of Science, Technology, and Environment in 1974. At present, the DOE is the main government agency responsible for monitoring and assessing violations of environmental regulations under the EQA 1974. Among the relevant responsibilities of the DOE are:

- to survey and inventory the pollution sources to be monitored
- to review, evaluate, and document effluent treatment technologies
- to develop and prepare regulations and guidelines.

The EQA requires inland water quality to be protected and enhanced for multiple beneficial uses conducive to public health, safety, and welfare. The DOE's programmes encompass a number of activities.

- *Establishment of river basin control regions.* Water pollution control of inland waters and management and allocation of water resources entail the establishment of water quality control regions in accordance with river basin drainage systems. A total of 49 river basin control regions have been established in peninsular Malaysia.
- *Water quality monitoring and surveillance.* In exercising regulatory functions, river monitoring and surveillance constitute the major activity in the water quality assessment programme. The water quality data are needed in order to determine the conditions of the rivers and assess the effectiveness of the instituted water pollution control measures.
- *Programme development.* The national water quality monitoring programme (1978) requires the routine water quality monitoring and surveillance of important river basins. Prior to this there was no systematic river water quality monitoring in the country except that carried out by the DID, which means mainly for drainage and agricultural irrigation. Before the programme was started, a network of monitoring stations at critical locations was established for major rivers, with respect to sources of public water supplies, river fishing, and agricultural irrigation. Each monitoring station is coded according to the system of the DID. This enables information on water quality and other hydrological data to be interchangeable between the DID and the DOE.
- *Water quality parameters.* The national waters are principally used for drinking, agricultural irrigation, fishing, and industrial processing, particularly for processing agro-based products such as palm oil, rubber, tapioca, sago, and pineapple. The basic parameters selected to represent the various qualitative aspects of the national waters are pH, biological oxygen demand (BOD), chemical oxygen demand (COD), suspended solids (SS), and ammonia nitrogen (NH_4-N). However, the kinds of pollutant sources found in a river basin will determine whether additional parameters, such as chloride concentration, conductivity etc., be analysed. Analysis is carried out by Department of Chemistry laboratories in Petaling Jaya, Penang, Ipoh, Johor Bahru, and Kuala Terengganu.

The DOE also undertakes studies to establish baseline information to tackle emerging pollution problems. The DOE is not empowered to stop or suspend the operations of factories and industries that violate the environmental regulations, except for rubber and palm-oil factories which obtained licences from the DOE; instead, the DOE has to seek the cooperation of the relevant local authorities to suspend the operations of factories or industries which violate the EQA. Amendments to the EQA have been made from time to time, including the mandatory use of environmental labelling, increases in the maximum fines to be imposed on various environmental polluters, and establishment of a fund to control and prevent oil spills.

There are a number of government departments responsible for ensuring that water quality remains adequate. The water-related functions of these departments are summarized below.

- Department of Irrigation and Drainage (DID) – hydrology subsection and surface water resource assessment.
- Department of Agriculture (DOA) – assess water requirements for crops.
- Department of Chemistry (DOC) – extend the services of water quality analysis to all water-related agencies.
- Department of Environment (DOE) – pollution control and water quality management, and planning, programming, implementing, operating, and evaluating matters pertaining to water resources.
- Geological Survey Department (GSD) – mapping and assessing groundwater resource potential.
- Mines Department (MD) – control water use for mining and water pollution through regulating discharge from mining activities.
- Ministry of Health (MOH) – carry out water supply programmes in rural areas and monitoring of microbial contamination.
- Public Works Department (PWD) – engage in planning of water supply, urban water supply schemes, regional water supply development, and rural water supply schemes.

Under the federal constitution, water supply within a state belongs to the authority of the state in terms of legislative and executive powers. Each state possesses a Water Supply Enactment or a Water Authority Enactment. The Water Supply Enactment provides that a state engineer may execute operations related to laying down of water mains and pipes for water supply in the area concerned. The state engineer is usually on the staff of the State Public Works Department (SPWD), of which the Water Supply Department is a subsidiary. The water supply section of the SPWD is responsible for water supply in the state. Only the state government of Selangor has a Water Works Department. The Kelantan Water Supply Department is privatized as Kelantan Water Sdn Bhd. The state agencies are engaged in two types of water supply; namely urban water supply and rural water supply. Water supply management is usually assisted by the federal PWD. Rural water supply by the state agencies is carried out in areas occupied by less than 10,000 residents. The MOH supplies water in areas where there are 200 to 500 residents and water supply facilities are not likely to be provided within the next five years.

Implementation and enforcement of the national environmental laws have often appeared weak and prone to loopholes. Since the establishment of those laws as regulatory instruments for environmental protection and pollution control, the legal framework has been largely restrictive. Regulatory instruments are known to perform relatively well in institutional compatibility and administrative feasibility, but not as well in terms of economic efficiency, cost-effectiveness, and enforcement. A number of ways are suggested to overcome these problems:

- introduce clear enforcement mechanisms at every stage in water quality management;
- rationalize and simplify water-related regulations;
- strengthen compliance monitoring of water pollution generators;
- make controls and sanctions for non-compliance with the water regulations more stringent by imposing tougher fines and penalties;
- simplify legal and administrative procedures for applying such sanctions;
- support regulations with economic instruments by applying financial incentives for compliance;
- improve information and publicity aimed at industries in respect to control requirements and pollution abatement technologies, and aimed at the public in respect to discharges and penalties for non-compliance;
- encourage the use of environmental auditing by the polluting industries as part of the Environmental Management System (EMS) ISO 14000.

International framework

The most relevant international negotiations on water resulted in the Marrakesh Agreement. The first World Water Forum was held in Marrakesh, Morocco, on 20–25 March 1997. At this Global Water Partnership (GWP) meeting, a proposal was submitted to the chairman of the Technical Advisory Committee to add an ecosystem sector to complement the existing ones dealing with water supply and sanitation, and irrigation and drainage. The Ramsar Convention on Wetlands of International Importance Especially as Water Fowl Habitat could become an associated programme within the ecosystem sector. The GWP has been interested in associating with the Ramsar Convention; Malaysia ratified this convention in March 1994 and Tasik Bera, an important wetland area, was added to the list of protected areas as a Ramsar site. If the Marrakesh World Water Forum agreements are incorporated into the Ramsar Convention, Malaysia will be obliged to accept the resolutions on water supply and sanitation, and irrigation and drainage, into its national and local regulations.

In order to control the transboundary movement of hazardous waste, Malaysia became a party to the Basel Convention on the Control of Transboundary Movements of Hazardous Waste and Their Disposal, effective from January 1994. Amendments were then made to the Customs Act 1967, the Customs (Prohibition of Import) Order 1988, and the Customs (Prohibition of Export) Order 1988 to include a list of chemical and other wastes to be controlled.

The general objectives of the national environmental policy are complemented and reinforced from time to time by bilateral and multilateral commitments through agreements, resolutions, and declarations such as:

- the Manila Declaration, 1981
- the Bangkok Declaration, 1984
- the Jakarta Resolutions, 1987
- the Manila Summit Declaration, 1987
- the Langkawi Declaration, 1989
- the Kuala Lumpur Accord on Environment and Development, 1990.

Apart from these, the Brundtland Report (Brundtland 1987) and the UN General Assembly Resolution 44/228, together with the events preceding the Rio Summit in 1992, have had a great impact in shaping the strategies for implementing policy objectives.

Water resource management programmes in Malaysia

In Malaysia, various water resources projects were undertaken during the Sixth Plan period to meet domestic and industrial demand and irrigation requirements. Efforts were also made to improve management and to ensure better distribution of water resources among various river basins to match supply and demand. By the end of the Sixth Plan period three new dams had been completed, bringing the total dams in operation to 72, with a total capacity of 25 billion cubic metres, while another three are under construction. Of these, about half were developed for water supply, 16 for multipurpose use, and the remainder for irrigation and hydro power.

In line with the need to provide safe drinking water, several urban and rural water supply programmes were implemented, with an emphasis on developing and upgrading source works, storage, and treatment plants as well as rehabilitating the distribution system. Efforts were also taken to enhance interstate water transfer, such as from the Sungai Muar in Johor to the Durian Tunggal dam in Melaka. Water transfer is also carried out from the Kelinchi dam in the Muar river basin to the Terip dam in the Linggi river basin in Negeri Sembilan. Measures were taken to improve the efficiency of the water supply system as well as water quality by reducing the rate of non-revenue water (NRW). Consequently, the national NRW rate decreased from 43 per cent in 1990 to 38 per cent in 1995. The percentage was expected to reduce further to 28 per cent in 2000 as a result of more efficient water supply systems.

The accelerated growth of several states requires a more systematic plan to be undertaken to coordinate the uneven distribution of water resources. This is done through the interstate water transfer projects. These projects involve the construction of various dams and associated water supply infrastructure. Apart from the source works projects, several new water supply projects, such as the construction of treatment plants and reservoirs for Kuala Terengganu, Kuantan, and Sungai Terip, are being be initiated. In line with the policy to increase accessibility of safe water in rural areas, the installation of new reticulation systems to distribute further water to rural households will be undertaken. About 3,600 projects, covering mostly isolated rural areas and benefiting about 262,400 households, will be implemented. The establishment of the National Water Resources Council (NWRC) will help resolve legal, institutional, and financial issues.

Privatization has been thought to be a vital part of the overall strategy to strengthen the role of private sector in the country's water resource development. This will contribute indirectly to the accelerated growth of the economy as well as assisting in the restructuring of objectives aimed at further enhancing *bumiputera* (ethnic Malays and other indigenous peoples) participation in the water resource corporate sector. The increased number of projects privatized during the Sixth Plan period reduced significantly the government's financial and administrative burden, thus enabling the government to allocate more resources to critical sectors of the economy.

Effectiveness of the governance structures

It can be seen from the previous discussion that there are a number of water-related laws for peninsular Malaysia and a number of federal and state agencies involved with water management. Historically and constitutionally, water resources have been managed by federal and state agencies. Some success has been achieved with water pollution control, particularly in palm-oil and rubber waste management, as can be seen under the prescribed premises procedure provided by the EQA. However, water pollution by organic matter remains a major problem. Despite having introduced the EQA almost three decades ago, many industries are still unable to comply with the regulations prescribed by the Act. There is a continuous need for stricter enforcement of the law against offenders by issuing compound notices, warnings, and notices to carry out remedial actions.

Although federal and state agencies are closely linked, state agencies dealing with state matters have to satisfy the state government requirements, which sometimes contradict federal opinions. This creates inefficiencies and sometimes delayed decision-making even on trivial matters. The government has acknowledged the need for better management of water resources, which resulted in the setting up of the National Water Council (NWC). On 29 June 1998, the NWC was renamed the National Water Resources Council. The NWRC is to coordinate the use and distribution of water among the states and also to integrate the management of land, forest, and water resources while ensuring uniformity in practices and law (see Figure 5.2). The NWRC will also function to establish a national water policy (NWP) to provide general guidelines which will become the principles for planning and managing 150 river basins. The secretariat of the NWRC is under the Works Ministry.

From the functions of the various government departments, agencies,

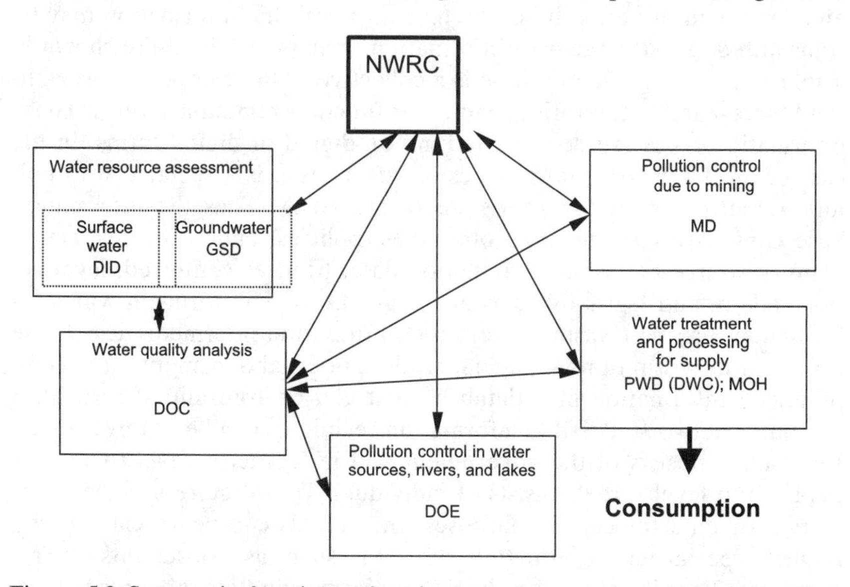

Figure 5.2 Systematic functions of various government departments, agencies, and ministries for water quality monitoring and supply

and ministries, it seems inevitable that water quality management is inefficient due to lack of coordination. This problem can be overcome if a coordinating body is set up by the government; for example, the NWRC can be the coordinating body in this case. The systematic coordination of the functions of the various government departments, agencies, and ministries is suggested in Figure 5.2. Clearly there are three main components of the system, namely the assessment component, the pollution control component, and the quality monitoring and supply component. These components can be coordinated by the NWRC through two-way interactions between the NWRC and the various departments and agencies. Water resource management guidelines and policies can be directed to the respective departments and agencies, while the feedback to the NWRC can be in the form of information, databases, and expert advice and suggestions. The NWRC is still in its infancy, and its exact role is yet to be determined.

Information availability and reliability

Information on drinking water contamination is not totally transparent and open to the public. Only certain types of information are accessible by the public and individuals. It is pertinent at this point to analyse and assess the types of water-related information that may be available from an agency's information base, the form in which the information may be transmitted, and the possible information recipients. These are shown in Table 5.6. The information base is a collective of information types gathered by research, monitoring, and other function-related activities. These information types are documented in non-digital or digital forms. In the case of the DOE, one of the types of information is associated with pollution control under the provisions of the EQA. Thus the information base consists of the agency's objectives, policies, and strategies, action plans, resource assessments, pollution data, licences conferred, legal actions taken, and legislative acts and regulations. The forms in which information may be transmitted are the information intermediaries. These can be in the form of newspapers, reports, internal documents, and computerized information in a database that can be transmitted through a computer network. Possible information recipients can be at three levels. Level one consists of the government and its agencies, level two is the public, and level three consists of individuals who require specific information or data for certain purposes. Individuals can be specialists who require specialized information or data, such as consultants or researchers. Thus it can be seen how the information flow takes place from the information base to the recipients via the intermediaries.

Table 5.6 Possible information flow

Information base of an agency	Information intermediaries	Possible recipients		
		Level 1	Level 2	Level 3
Objectives, policies, and strategy statements	Documents and IT (internet based)	Central government, federal and state agencies	Public	Specialized individuals
Corporate documents	Annual reports			
Action plans	Planning reports			
Resource assessments	Assessment reports			
Raw data from research and monitoring	Computer database	–	–	–
Analysed data	Computer database	State departments, federal and state agencies	Public	Specialized individuals
Summarized data	Annual reports and IT			
Acts	Published documents and books			
Legal actions taken	Informal documents and mass media			
Licences	Internal documents	–	–	–

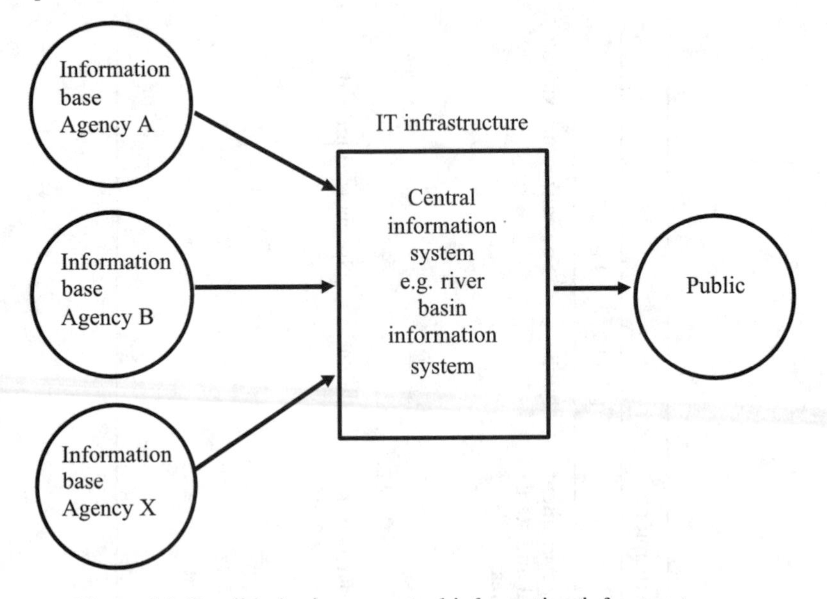

Figure 5.3 Possible basic conceptual information infrastructure

The present information infrastructure of Malaysia relies on the internet through the joint advanced research integrated networking (JARING) system, which uses nodes located in 20 major towns around the country. Having adequate facilities for the efficient flow of information, a concerted effort is needed to develop the information system for better management of the water resources of the country. Dissemination of water-related information may be improved by setting up a river basin information system, as shown in Figure 5.3. This is because access to information that is current and accurate is crucial when it involves the management of land, forest, and water resources. To improve resource management, public agencies must work together to develop integrated databases on land use, forests, rivers, lakes, groundwater, and other vital resources. Information technology holds the key to overcoming the problems of work among state and federal agencies while saving time in data retrieval.

Information technology (IT) also allows some room for transparency and accountability. With the right IT application, obtaining important data involving water can be carried out quickly, and the extent of environmental degradation can be detected and action taken promptly. The creation of a river basin information system is crucial. To integrate river basin management, the current practice of collecting and collating data must change. Ways must be found to reconcile the different formats of information, as some are in numerical form while others are maps and structural plans. Most of the existing databases are for the exclusive use of the respective government agencies. This discourages information sharing, giving rise to the common problem of presenting an incomplete picture of the situation. The federal-state dichotomy over natural resources also hinders prompt decision-making. Data on river basins are largely in the hands of the federal government; nevertheless, the state governments decide on the fate of rivers as it pertains to land use, and most of the information stored at the state level is still in non-digitized forms.

Role of NGOs

NGOs in Malaysia can be divided into three categories depending on their activities and the area they are involved in: international NGOs (INGOs) and their domestic affiliates; NGOs totally focused on domestic issues; and knowledge-based NGOs working as think-tanks, seeking to influence policy and propose solutions, or working as research bodies publishing science or policy research. There are a number of environment-related NGOs in Malaysia. The establishment of these NGOs is governed by the rules and regulations of the Registrar of Society (ROS) of Malaysia. The best-known NGOs are:

- Malaysian Nature Society
- Friends of the Earth Malaysia (Sahabat Alam Malaysia)
- Environmental Protection Society of Malaysia
- Worldwide Fund for Nature (WWF), Malaysia.

They play an increasingly important role in all aspects of environment and development, including issues pertaining to water. Nevertheless, their involvement is not specific to water only. Generally, NGOs consist of personnel with expert knowledge and ideas who specialize in the issues being negotiated. The people involved are also dedicated to the issues, transcending economic or class interests and sectoral boundaries. NGOs have domestic political clout, representing interests that are not otherwise represented, attracting attention, and influencing some if not all political processes. The roles played by NGOs are many, and may include:

- agenda setting – defining a new issue, or redefining an old issue
- lobbying government to accept a position, general action such as consumer boycotts, propaganda, and education programmes, and legal suits

- drafting alternative text for a convention, usually in advance of the convention negotiations
- lobbying negotiators
- monitoring and reporting on implementation.

All these roles can be played by environmental NGOs in issues relating to water in Malaysia. As an example, the Federal Territory Counselling and Service Centre (FTCSC) submitted a 30-point memorandum to the Selangor state government on ways of solving the water problem in the Klang valley. The FTCSC was the first NGO to come forward with suggestions in the 1998 water crisis.

NGOs have worked hand in hand with the government, especially in carrying out environmental awareness, education, and training programmes. They also took part in the "Love Our Rivers" awareness campaign (Malaysia 1996).

Impacts of Malaysian industry on water resources

Some of the main problems of industry polluting fresh water and drinking water are the production of environmental pollutants, improper management of treatment systems, increased milling capacity, malodour, and others. The various problems, the major pollutant sources, and their locations will be discussed according to the respective industries. An overall picture of the industrial impact on water resources is summarized in Table 5.7.

In general, domestic sewage, industrial effluent, and animal husbandry effluent can be considered as the main pollution sources to river water quality. Effluent from mines also causes pollution in the form of suspended solids (SS) due to its large content of particles. Densely populated towns are the main pollution sources because of the large amount of domestic sewage. On the other hand, domestic sewage in rural areas does not affect river water quality, even though it has the possibility of causing local pollution. This is because the population density is still low, and domestic sewage is not being discharged in concentrated amounts into any particular watercourse. It can be noted that most major towns are located in coastal regions.

Most of the rubber factories can be found in the northern region, followed by the central, southern, and eastern regions. Therefore, it is expected that these regions show higher effects of pollution from rubber effluence. Complaints were mainly concerned with malodour, even though efforts were taken by the DOE to ensure that factories practise good housekeeping and install proper control equipment. Some factories were taken to court under Section 16(1) of the EQA 1974 for not com-

Table 5.7 Industries impacting water resources

Types of industries	Problems
Agro-based industries • Crude palm oil • Raw natural rubber processing	Improper management of treatment systems, increased milling capacity, malodour, and problems related to the disposal of wastes
Wood-based industries • Sawmills	Sawdust discharged into rivers causing water pollution and sedimentation, indiscriminate disposal of wood waste, inefficient burning of wood waste and unapproved incineration
Manufacturing industries • Rubber-based industries • Textile industries	Sewage and industrial effluent, lack of proper effluent treatment systems
Food manufacturing industries • Fish mills • Soft drinks	Absence of proper wastewater treatment systems, under-capacity of the existing treatment system, sewage and industrial effluents
Others • Tin mining • Pig farming • Oil refinery • Densely populated towns	Large amounts of suspended solid that do not settle readily, piggery waste discharged into rivers, industrial effluent affecting river quality, domestic sewage discharged into rivers

plying with licensing conditions. As a whole, the status of compliance of the raw natural rubber factories looks very promising, and they are moving towards achieving 100 per cent compliance. On the other hand, other rubber-based industries continue to pose problems with regard to disposal of vulcanized rubber wastes.

The effluent from palm-oil processing can generally be described as a highly viscous brown liquid, high in total solids and oil, and having BOD and COD values approaching 22,000 mg/l and 61,000 mg/l respectively. The states with the highest number of palm-oil mills are Perak, Selangor, Johor, and Pahang. The effluent from rubber factories and palm-oil mills has a high value for organic matter content. They are controlled by the DOE under the Environmental Quality (Prescribed Premises) (Raw Natural Rubber) Regulation 1978 and the Environmental Quality (Prescribed Premises) (Crude Palm Oil) Regulation (Environmental Quality Act 1974).

Most animal husbandry effluence comes from pig farming. The pig population is centred in particular areas – for example Penang, Perak, Selangor, Negeri Sembilan, Kota Kinabalu, Kudat, Sandakan, Tawau, Tuaran, Kucing, Sibu, and Miri – and is on the rise. As pig farms usually have no pond or space for pig-waste treatment, piggery waste is becoming a major source of water pollution. The lack of space in lagoons hinders their implementation of purification of pig waste. Hence, waste from pig farms is usually discharged to rivers without being purified.

The food manufacturing industries also contribute significantly to water pollution in the country. Their non-compliance is largely due to the absence of proper wastewater treatment systems, undercapacity of existing treatment systems to cater for the increased production of the industry, and lack of maintenance of wastewater treatment systems. Similarly, indiscriminate disposal of wood wastes from wood-based industries is of great concern to the environment, causing not only the pollution of air, water, and land, but also affecting the people and properties around the milling operations. The DOE has encouraged the reutilization of wood wastes generated in the wood-processing mills as fuel for kiln-drying operations, steam boilers, brick manufacturing, charcoal making, manufacturing of particle board etc. In spite of this, the problem still prevails and is worsened by the burning of wood wastes in inefficient and unapproved incinerators.

Many of the textile industries in the country are still operating without proper effluent treatment systems. As a result, a few places in the states of Perak, Johor, Selangor, Kelantan, and Terengganu face serious river water pollution. For industries that have wastewater treatment systems, however, compliance with the stipulated discharge standards of the Environment Quality (Sewage and Industrial Effluents) Regulation 1979 remains a problem, especially with respect to the parameter for COD.

Enforcement of the existing environmental laws and regulations is essential. It has been stepped up to ensure the capability of the industrial sector, and in particular to control the production of environmental pollutants and to practise effective storage and disposal methods. The enforcement of regulations for the management of scheduled wastes has been intensified, and the number of waste generators complying with the requirements of the regulations increased significantly through the notification and assessment scheme and licensing mechanism. However, activities in illegal dumping of toxic wastes still persist throughout the country, warranting serious enforcement action.

The implementation and enforcement of the mandatory environmental impact assessment (EIA) procedure and requirements under the Environmental Quality (Prescribed Activities) (Environmental Impact Assessment) Order 1987 have been stepped up. In an effort to improve current enforcement programmes, work has resumed to review various environmental laws and regulations within and outside the jurisdiction of the DOE and to formulate new regulations. The establishment of a telephone hotline in the Selangor DOE, judging by the great response from the public, has given extra effectiveness to routine enforcement activities. The feedback received has helped the department in combating environmental pollution in a more efficient manner. Cooperation of the Air Police Traffic Unit has been sought to carry out aerial surveillance of environmental pollution pertaining to illegal dumping of toxic wastes.

Public awareness and involvement as part of governance

Public awareness about the deterioration of the environment in ways that affect the quality and quantity of water in Malaysia is growing steadily. For example, in the recent past many river systems near urban and suburban areas have been polluted by household materials and solid wastes. This situation has improved recently except in slum areas, where improper waste management still continues. Polluting rivers with household solid waste resulted from the public's lack of a caring attitude and ignorance of the importance of a clean environment for living – a typical scenario in a survival society. The growing awareness of and concern for the environment have resulted from easy access to information through the mass media, education at various levels of the society, and enforcement of the more stringent environmental regulations. A more important aspect with respect to the growing awareness of and concern for the environment is the elevation of the standard of living with economic development.

Public opinion has not been an important determining factor in the development of federal and state government policies pertaining to any environmental issue in relation to water resource management. Nevertheless, it is a normal practice for the government to conduct public opinion surveys through questionnaires for gauging public feelings, perceptions, or behaviour before any project development, including those of water resources, is undertaken. This can be regarded as a form of public participation in a narrow sense because it normally has several dimensions due to the different roles citizens play in society, such as workers, consumers, and polluters. But it has not gone further – towards, for example, the creation of citizen advisory committees. Such committees can provide additional insight about the extent of a given environmental problem, as well as the social and political implications if the problem is left uncorrected. However, effective public participation requires that the government is responsive to the views of the public.

Nothing can be more damaging to public confidence when carrying out new governmental strategies than a perception that the government has never responded impartially to public participation in the process.

The management of water resources that may attract public interest is usually at the local level, with less appreciation of the long-term needs at regional or national levels. More often than not cost is the main criterion for choosing management options. If the public through its participation can be persuaded of the severity of a problem (e.g. water pollution), it can appreciate more easily the need for immediate environmental control programmes which are not necessarily the least expensive. Thus, public participation can be an important component of effective pollution control programmes.

REFERENCES

Aminuddin, B. Y., M. L. Sharma, and I. R. Willet (eds). 1996. *Agricultural Impacts on Groundwater Quality*. Proceedings of an international workshop held in Kota Bharu, Kelantan, Malaysia, 24–27 October 1994. Canberra: ACIAR Proceedings No. 61.

Brundtland, G. (ed.). 1987. *Our Common Future: The World Commission on Environment and Development*. Oxford: Oxford University Press.

Environmental Quality Act. 1974. Environmental (Prescribed Premises) (Crude Palm Oil) Regulations 1977. Government of Malaysia.

Environmental Quality Act. 1974. Environmental Quality (Prescribed Premises) (Raw Natural Rubber) (Amendment) Regulations 1980. Government of Malaysia.

JICA. 1982. *JICA National Water Resources Study, Malaysia*. Tokyo: Japan International Cooperation Agency.

Malaysia. 1986. *Water Quality Criteria and Standards for Malaysia: Final Report*. Kuala Lumpur: Ministry of Science, Technology, and Environment.

Malaysia. 1991. *Sixth Malaysia Plan 1991–1995*. Kuala Lumpur: Government of Malaysia.

Malaysia. 1996. *Seventh Malaysia Plan 1996–2000*. Kuala Lumpur: Government of Malaysia.

6

Water – The lifeline in Thailand

Monthip Sriratana Tabucanon

Thailand has achieved an exceptional record of economic development over the last 30 years, as witnessed by the rapid expansion of the national economy at an average rate of 7.8 per cent per annum. Thailand has undergone considerable industrial development and urbanization, but the process of urbanization and industrialization has led to a dramatic increase in the burden placed on sewerage infrastructure, waste treatment facilities, and the high demand for water supply.

Until recently, water resources in Thailand have been abundant and in excess of the need for water utilization. Development in the past has gone full steam ahead. But since the country has become more industrialized and the population has increased rapidly, the need for water has grown while the amount of water available has remained roughly constant. This has given rise to problems resulting from conflicts over water use between various sectors, such as agriculture, domestic consumption, hydro power, flood protection, and inland waterway transportation. Another noticeable problem is that nearly all the suitable sites are either in the heavily populated areas or in the national park reserves. This poses difficulties for the development of large-scale projects, in terms of either water resource development or wastewater treatment facilities.

Thailand's water resources

Status of water resources development in Thailand

Thailand has a total area of 513,115 square kilometres. The average annual rainfall is about 1,550 mm, and the average volume of rainfall is about 800 billion cubic metres annually. However, the rainfall is not evenly distributed throughout the country. The southern region receives the highest rainfall while the northern and central regions receive the lowest. Thailand now has 26 large dams and reservoirs with a capacity of 66 billion cubic metres and effective storage capacity of about 43 billion cubic metres. The potential to develop additional large dams and reservoirs in Thailand is now very limited due to physical limitations and growing concerns regarding their environmental impacts. The Royal Irrigation Department, Ministry of Agriculture and Cooperatives is the agency responsible for irrigation management. Table 6.1 shows the distribution of large reservoirs and dams and their water storage in the country.

The most important and the largest river basin in Thailand is the Chao Phraya basin, the area with the highest concentration of population and economic activities in the country. It originates in the mountain ranges in the north and covers nearly all the areas in the northern and central regions. The average annual run-off at the river mouth is 30,300 million cubic metres, or 170 mm in terms of depth. At the end of the Eighth National Plan, a pilot project was implemented on the use of the watershed system approach in water resource management of the Chao Phraya basin.

The second important river basin is the Mae Klong basin. The two main tributaries, the Quae Yai and the Quae Noi, originate in the mountain

Table 6.1 Reservoirs and water available in selected years, Thailand (million m^3)

Region	Number of reservoirs	Storage capacity	Effective storage capacity	Actual usable water		
				1990	1993	1995
North	5	23,612	16,934	8,603	5,629	13,220
North-east	9	7,092	5,550	3,572	2,706	3,285
West	6	28,160	14,708	7,369	5,046	10,393
East	4	204	184	102	105	106
South	2	7,043	2,431	1,986	371	2,224
Total	26	66,111	42,807	21,632	14,857	29,228

Source: RID (1995)

ranges in the west, near the Myanmar border, and join together to form the Mae Klong river some 80 kilometres from the Gulf of Thailand. The drainage area is 33,000 square kilometres. The average annual run-off is 13,400 million cubic metres, or 406 mm of depth. In the other regions, the major river basins are the Bang Pakong basin in the east, and the Maekhong and Salawin basins in the south.

In order to conserve the water resources, the need for watershed classification has been recognized in Thailand for more than two decades. In October 1979 a Committee on Watershed Classification was officially set up. A major objective of watershed classification projects is to formulate land-use plans for the conservation of natural resources, and in particular water resources, with a view to their sustainable use. The classification takes into consideration physical characteristics, including stable features such as land form, geology, soil, elevation, and slope. Forest cover and environmental features of landscape units, which are less stable and interact with climatic trends and human uses, are also considered. Land areas of one square kilometre in size are selected and assigned a numerical watershed class value based on characteristics of their features. Values for each variable are read either from topographic, soil, geology, or forest map information for each cell. The theoretical modelling techniques have highlighted the importance of watershed management in Thailand. The watershed classification study was completed and the Cabinet approved its implementation in all the watershed regions of the country in 1995. Table 6.2 shows the watershed classification (WSC) and recommended land use in Thailand.

The surface water resource assessment in the 25 watershed areas of the country was completed in 1992. Topographical characteristics and the potential of water resource development in each watershed were analysed. The assessment of groundwater nationwide was conducted by the Department of Mineral Resources. At present, 1:500,000 and 1:1,000,000 mapping of water resources are completed, with detailed 1:100,000 scale maps of provincial groundwater resources in the north-east also completed. The detailed groundwater resource maps for other provinces are currently being prepared.

Groundwater resources exist throughout the country. The quantity and quality of groundwater vary according to local hydrogeological conditions. Usually, large and high-yielding aquifers occur in alluvium and terrace deposits, and groundwater also exists within rock formations in limestone, sandstone, and some types of shale. Under the national potable water scheme, a total of 22,361 serviceable wells have been drilled. It is estimated that about 880 million cubic metres of groundwater was abstracted in 1998. Of that amount, approximately 629 million cubic metres was used for domestic and industrial water supply in Bangkok and its

Table 6.2 Watershed classification and recommended land use in Thailand

WSC class	Characteristics and major land uses recommended	Watersheds approved by Cabinet
WSC 1A	Protected forest and headwater source areas at higher elevations and steep slopes. Primary headwater areas should remain under permanent forest cover	Ping-Wang: 28 May 1985 Yom-Nan: 21 October 1996
WSC 1B	Similar to WSC 1A but some areas cleared for agricultural use or occupied by villages. Primary headwater areas require special WSC measures, replanted to forest or maintained in permanent agro-forestry	Mun-Chi: 12 July 1998
WSC 2	Protected and/or commercial forests at higher elevations with steep slopes. Land forms less erosive than WSC 1A or WSC 1B. Secondary headwater areas may be used for grazing or certain crops with WSC measures	Southern: 7 November 1989 Eastern: 18 November 1991
WSC 3	Uplands with steep slopes and less erosive land forms. Areas may be used for commercial forests, mining, grazing, fruit trees, or certain agricultural crops with WSC measures	Western-Central-Pasak, Northern, and North-eastern Border: 21 February 1995
WSC 4	Gently sloping lands suitable for row crops, fruit trees, and grazing with moderate need for WSC measures	
WSC 5	Gentle to flat areas used for paddy fields or other agricultural uses with few restrictions	

vicinity. The budget for groundwater development during 1998 totalled 1.4 billion baht.

Groundwater has a high probability of being polluted by some industries. The 1997 groundwater quality survey by the Environmental Research and Training Center (ERTC) reported that volatile organic compounds and arsenic are the main problems of pollutants in the northern and southern part of Thailand respectively. According to a joint

United Nations University (UNU) and ERTC study, the concentration of volatile organic compounds (VOCs) in the groundwater in Thailand is still within the WHO standard.

The coastal and marine environment of the Gulf of Thailand is degraded by pollutants from both land-based and maritime sources. Land-based sources, most of them domestic, contribute 70 per cent of marine pollution. It is estimated that more than 200,000 tonnes of contaminants (measured as biological oxygen demand – BOD) are discharged into the Gulf annually. Studies over the last decade reveal that trace metals, petroleum hydrocarbons, and organochlorine pesticides in seawater, sediment, and aquatic tissue tend to be higher in the area of the inner Gulf and at the river mouths. However, these levels are still within the coastal water quality standards of Thailand.

Key issues

Protection of water resources, water quality, and aquatic ecosystems

Domestic sewage, industrial effluents, and agricultural run-off all contribute to pollution of surface water and groundwater in Thailand. The 1997 water quality survey by the Pollution Control Department (PCD) reported that 14 per cent of surface water resources are suitable for aquatic animals and general human consumption, another 49 per cent are usable for agriculture and general consumption, and the remaining 37 per cent are of poor quality. Water pollution is most severe in the Bangkok Metropolitan Region, especially in the lower reaches of the Chao Phraya and lower Tha Chin rivers, due to the domestic and industrial activities. During 1991 and 1997, the average level of dissolved oxygen (DO) in the middle reaches of the Chao Phraya, Mae Klong, and Bang Pakong rivers was higher than the standard level of 4 mg/l.

High concentrations of total coliform bacteria (TCB) represent a significant health risk for communities dependent on untreated river water for cooking and drinking. Between 1991 and 1995, the standard of 20,000 MPN/100 ml was often exceeded by a factor of 10 or more in Thailand's four major rivers.

In general, the Ministry of Agriculture and Cooperatives has the main responsibility for protection of water resources; the Royal Forest Department is responsible for all protected forest areas and coastal resources that are vital for water resources protection; the Royal Irrigation Department is responsible for most irrigation water in the country; the

Table 6.3 Wastewater treatment systems initiated by the BMA

Project name	Capacity (m^3/d)	Cost (US$ million)	Status
Si Phraya	30,000	11	Completed in 1994
Rattanakosin	40,000	21	Completed in 1999
Collective wastewater treatment system	350,000	154	Completed in 1999
Yan Nawa	200,000	110	Completed in 1999
Rach Burana-Nong Kham-Phasi Charoen	65,000 and 157,000	118	Completed in 2001
Collective wastewater treatment system Phase IV	150,000	95	Completed in 2001
Thonburi wastewater treatment system	NA	250	Completed in 2003

Source: BMA, unpublished data (1999)

Department of Fisheries is responsible for freshwater and seawater fisheries and aquaculture; and the Department of Land Development takes care of land and soil conservation. The Ministry of Science, Technology, and Environment, Ministry of Public Health, and Ministry of the Interior house agencies which are mainly responsible for monitoring and control of water quality.

Over the past few years, the government has made major investments in constructing wastewater treatment systems. More than US$1,200 million has been allocated for wastewater and sewage projects for the Bangkok Metropolitan Area (BMA) and its vicinity. Similarly, more than US$400 million has also been allocated to projects for municipal and sanitary district areas. Table 6.3 shows the cost of wastewater treatment systems in the BMA.

Thailand has 145 municipal areas and 995 sanitary district areas. About 34 areas are served by wastewater treatment systems and 33 more systems are being constructed. The Wastewater Management Agency (WMA) has been established, affiliated with the Ministry of Science, Technology, and Environment, to handle wastewater collection and treatment nationwide. At present, Pattaya and Pha Tong sanitary districts, Phuket are the only service areas when sewage fees can be collected.

In 1996 there were 98,977 factories registered with Department of Industrial Work (DIW); among these 9,831 plants were classified as factories causing water pollution. The wastewater originating from factories comes under the responsibilities of Industrial Estate Authority of Thailand (IEAT); Table 6.4 shows the industrial estate wastewater facilities.

Drinking water supply and sanitation

Thailand defines availability of drinking water supply as the ability of households to gain access to safe drinking water within one kilometre from their village. Thailand has long since developed strategies to solve drinking water supply problems by building village water storage tanks, providing groundwater access to villages, and building rural piped-water systems. Rural households are encouraged to store rainwater for domestic use. Drinking water problems usually follow extended droughts, as was the case in 1992 when more than 28,000 villages suffered from drought and acute shortages of drinking water.

Several government agencies are involved in domestic and drinking water development in rural Thailand. These include the Office of Accelerated Rural Development, the Department of Community Development, the Department of Mineral Resources, and the Department of Health. Development of drinking water supply in Thailand has progressed satisfactorily. The most important source of clean drinking water for many villages is groundwater. Thailand provided more than 150,000 deep and shallow wells for clean water supply to rural areas, as shown in Table 6.5. In 1994 more than 80 per cent of all villages had access to year-round water supply, and more than 70 per cent of villages had access to clean drinking water. When monitored by household, over 86 per cent of rural households had access to clean drinking water.

To ensure good sanitation, the government of Thailand regularly monitors and assesses the quality of drinking water supplies. Data on raw water quality entering and leaving the treatment plants supplying water to Bangkok are provided in Table 6.6. For intake water, the standards are those associated with Class 3 water. For water flowing from the treatment plant, the standards are for drinking water. Considering the raw water quality at the plants in 1997, coliform counts are clearly in excess of standards, as expected.

Direct access to drinking water supply for households in rural areas has improved substantially over the past decade. However, Thailand has not yet met its target of 95 per cent of all households having access to at least 55 litres per person per day of clean drinking water. More efforts have to be made, especially in areas where frequent and severe droughts occur, for Thailand to expand the availability of sanitary drinking water supply to rural households to meet this target.

Water and sustainable urban development

About 78 per cent of the population in Bangkok has access to municipal water supply, while the remaining 22 per cent still have to rely on

Table 6.4 Industrial estate wastewater facilities

Industrial estate	Zone/province	Maximum design wastewater capacity	Maximum BOD content capacity	Maximum BOD loading	Wastewater treatment system
Amata City	3/Rayong	22,000	NA	NA	NA
Bang Chan	1/Bangkok	10,000	NA	NA	NA
Bangpa-In	2/Ayudhaya	13,000	500	6,500	AS
Bangpakong	2/Chonburi	11,500	500	5,750	AL
Bangpice	1/Samut Prakarn	5,000	500	2,500	AS
Bangpoo	1/Samut Prakarn	2,300/2,000/3,600	1,000/500/500	23,000/1,000/1,800	AL/RBC/AS
Chonburi	2/Chonburi	13,460	750	10,095	AS
Eastern	3/Rayong	12,000	750	9,000	AS
Eastern Seaboard	3/Rayong	37,000	NA	NA	NA
Gateway City	2/Chachoengsao	37,229	500	18,615	AS
Gemopolis	1/Bangkok	11,200	250	2,800	AS
Hitech	2/Ayudhaya	16,800	500	8,400	AS
Ladkrabang	1/Bangkok	15,800	1,000	15,800	AS
Laemchabang	2/Chonburi	30,700	500	15,350	AS
Map-Ta-Phut	3/Rayong	8,000	500	4,000	AS
Nong Khae	2/Saraburi	14,400	500	7,200	AS
Northern Region	3/Lumphun	5,600	750	4,200	AL
Padaeng	3/3/Rayong	NA	NA	NA	NA
Piothi	3/Pichit	5,100	NA	NA	NA
Pinthong	2/Chonburi	1,200	NA	NA	NA
Ratchaburi	2/Ratchaburi	10,000	NA	NA	NA
Saharatana Nakorn	2/Ayudhaya	6,400	500	3,200	AS
Samut Sakhon	1/Samut Sakhon	21,000	500	10,500	AS

Saraburi	2/Saraburi	8,800	500	4,400	AS
Southern	3/Songkla	3,000	NA	NA	NA
WellGrow	2/Chachoengsao	14,400	500	7,200	AL/AS

Source: IEAT (1999)
NA = not applicable
AS = activated sludge
AL = aerated lagoons
RBC = rotating biological contactors

Table 6.5 Development of groundwater supply by major agencies in Thailand (number of wells)

Agency	End of the Sixth Plan (1991)	Seventh Plan (1992–1996)	Eighth Plan target (1997–2001)
Department of Mineral Resources	58,126*	10,813**	23,500
Office of Accelerated Rural Development	27,922	21,900	12,750
Department of Civil Service	16,713	17,920	20,000
Department of Health	14,456	13,300	13,000

Source: NESDB (1997)
* Up to 1994
** For 1995 and 1996

groundwater sources. The main body responsible for municipal water supply in Bangkok is the Metropolitan Waterworks Authority (MWA). Regional cities are served by the Provincial Waterworks Authority (PWA), and many of the local administrations in provincial municipalities also operate water supply services to their communities.

In coordination with the Seventh National Plan, the MWA operated under the guidance of its master plan (1992–1999). Over this period, the MWA expanded its pipelines to cover another 3,000 kilometres, especially in the east and the west of Bangkok where groundwater extraction has been excessive. Two new raw water supply canals, with a total length of 100 kilometres, will be built to support water production plants. By 1997, all groundwater extraction in the critical land subsidence zones was banned.

The PWA has also rapidly developed its piped-water supply system in the Bangkok vicinity to support the substitution of groundwater extraction. To accelerate this, the PWA implemented a joint-venture project with the private sector to invest more than 2,000 million baht in a municipal water supply system to the north of Bangkok. It is also preparing to expand water supply to the south and the west of Bangkok with total investments of more than 3,400 million baht, again using the joint-venture approach.

Thus, Thailand continues its effort to extend water supply to urban areas to support domestic and economic activities. Private participation in providing public utilities was developed to improve the efficiency and capabilities of the service. Nevertheless, some major problems persist in urban water systems:

Table 6.6 Raw water quality at MWA treatment plants, 1997

Parameter	Unit	Bangkhen		Samsen		Standard (maximum)
		Maximum	Minimum	Maximum	Minimum	
Physical-chemical parameters						
Colour	Pt-Co	17.0	6.0	18.0	5.0	–
PH	pH units	7.81	7.15	8.0	6.6	5.0–8.0
Total solids	mg/l	854	165	258	150	–
Dissolved oxygen	mg/l	6.0	3.8	6.9	4.5	4.0
Free chlorine	mg/l	3.2	0.5	2.8	0.3	2.0
Nitrate	mg/l	2.0	0.0	0.5	0.0	5.0
Manganese	mg/l	0.9	0.1	0.2	0.01	1.0
Lead	mg/l	0.7	0.01	0.03	0.01	0.06
Bacterial parameters						
Coliform bacteria	col/100 ml	236,000	12,000	82,000	10,000	20,000
Faecal coliform bacteria	col/100 ml	1,880	600	2,460	220	4,000

Source: Metropolitan Waterworks Authority (1997)

- insufficient water resources, especially during the dry season;
- the need to extend the water supply system to areas now using groundwater;
- high water losses;
- pricing of municipal water supply to cover cost and encourage conservation.

Water for sustainable food production and rural development

Agriculture in Thailand relies on either rainfall or natural flow and, where available, on irrigation water. Due to natural resource and environmental constraints, the potential to expand irrigation is increasingly limited. Thailand developed several complementary measures to sustain agricultural production and rural development, including a shift of emphasis from large- to medium- and small-scale irrigation, crop diversification, and increased efficiency in water use.

Crop diversification and restructured cropping were strongly supported to convert water-intensive agriculture to water-saving agriculture. The Thai government launched a five-year project (1993–1997) to encourage farmers in the Chao Phraya plain to convert from a second rice crop to other economic crops that need less water. With a budget of 3.2 billion baht (US$1 is around 43 baht), the project covered 1.1 million rai of irrigated land in 22 provinces (1 hectare is 6.25 rai, thus the project covered 176,000 ha). In 1994, the project converted about 333,000 rai (53,000 ha) of second-crop rice to these other, more water-saving, economic crops. During the Seventh National Plan, a three-year programme (1994–1996) to diversify from monocropping to multiple farming was also launched. Farmers were encouraged through various incentives to switch from rice, cassava, coffee, and pepper to other economic crops.

Thailand gradually introduced more stringent control of water use, especially for non-agricultural sectors. Water fees are now charged for non-agricultural uses, including water intake for municipal water supplies. Although water charges in agriculture were proposed in the Seventh National Plan, social and economic conditions did not permit their implementation. Increased efficiency in agricultural water use has mostly occurred through incentives to switch crops. It is foreseen that the new watershed system approach and the proposed water resource law will restructure the water allocation system among different groups, different economic sectors, and different locations or watersheds. The new dimension of water resource management will certainly have substantial implications for water resource value, and hence lead to better allocation of this resource.

Legal and institutional framework

Legal framework

Water law is probably the most important facet of water resources development and management. In the absence of a legal framework, resource management cannot be implemented without confusion and conflict. However, legislation directly related to water resources management in Thailand has been limited. Some of the existing laws that are related to surface water management are the People's Irrigation Act 1941; the Public's Irrigation Act 1942 (Fourth Amendment 1975); the Ditch and Dike Act 1962; the Land Consolidation for Agriculture Act 1974; and the Land Reform for Agriculture Act 1975. In none of these laws is there a clause that clearly defines the rights and duties of irrigation users, or the authority and duties of irrigators. Due to the increasing water demand by various sectors, the lack of a law which clearly defines the rights and duties of riparian users may lead to future conflicts resulting from competitive uses of water.

Groundwater Act 1977

The Groundwater Act 1977 empowers the governmental control of groundwater development and management. Under the provisions of the Act, one must obtain an official permit to utilize groundwater from designated groundwater areas. The Act governs drilling and the use of groundwater, as well as the disposal of wastewater into an aquifer through a well. At present, the Act is being implemented in some areas where overexploitation and groundwater quality have been critical problems. The Director-General of the Department of Mineral Resources (DMR) is the chief executive administering the Act, under the direction of the Minister of Industry.

The Ministry of Industry has issued directives on the technical principles of groundwater management.

- Bangkok and the five adjacent provinces have been designated as the Bangkok groundwater area. Groundwater occurring at depths exceeding 15 metres below the surface in this area is subject to control under the Act.
- Specifications for drilling and the construction of wells are provided under the Act. Standard forms for daily drilling reports, well records, and other information are prescribed.
- Methods of groundwater extraction and conservation are outlined.
- Technical measures to protect groundwater from pollution are described and drinking standards have been issued.

- Technical principles are given for the disposal or injection of water or liquids into an aquifer through a well. Emphasis is placed on the monitoring of water quality in the affected aquifer and neighbouring aquifers through a network of observation wells.

Those who violate the Act or ministerial directives are punishable by fine or imprisonment. It is worth noting that the area designated as the groundwater area only covers Bangkok and adjacent provinces, with a total area of 700 square kilometres. Though land subsidence and salt-water encroachment problems due to groundwater pumping in Bangkok are well recognized, drilling and pumping of groundwater cannot be totally eliminated. This is, in part, due to the inability of the MWA to meet water demand from other sources. Nevertheless, the implementation of the Act serves to reduce the severity of the problem.

Public Health Act 1968, 1978, 1988

The Act is comprehensive, regulating the disposal of rubbish, filth, and dirt as well as authorizing local authorities to issue by-laws or rules stipulating the methods and procedures to be used in such disposal. The local authorities are also empowered to control commercial undertakings which are likely to be injurious to health, unsanitary dwelling places, latrines, night-soil receptacles, urinals, and other sites, facilities, or watercourses likely to represent a hazard to health. The Act also prescribes penalties. Under the Public Health Act, drinking water quality standards and bottled drinking water quality standards have been set up.

Act for Cleanliness and Tidiness of the Country 1960

This Act regulates and controls public offences, including disturbance and anti-aesthetic activities. It specifically prohibits the dumping of waste into rivers or canals.

Building Control Act 1979

This Act and its municipal by-law of 1979 updated an earlier Act and by-law of 1936 and 1940 respectively. The local authority is empowered to issue by-laws controlling the number and type of bathrooms and toilets. It can also control storm water and wastewater drainage under the Act.

Factories Act 1978, 1982

This Act imposes certain duties on industrial concerns using processes which will lead to the discharge of defined levels of effluent.

National Environmental Quality Act 1975, 1992

The National Environmental Quality Act, amended in 1978, created the National Environmental Board (NEB). It authorizes the NEB to perform functions which are mostly concerned with policy development and coordination with other government agencies in matters relating to environmental quality. The 1978 amendment empowers the NEB to issue ministerial regulations covering certain designated projects for which the submission of an environmental impact assessment (EIA) is a precondition of NEB approval. Several committees have been formed under the NEB umbrella; those responsible for various aspects of hazardous waste management include the Committee on Water Quality.

Targets

In order that environments and natural resources are maintained in a condition in which they are of lasting benefit to the quality of life of the Thai people and to national development, the Eighth Plan sets out to ensure that the utilization of water resources is counterbalanced by rehabilitation and protection programmes and to promote more effective water management, involving the collaboration of various different sectors of society. This approach will help achieve greater balance in ecosystems and environments. Opportunities will be provided for local people and organizations to play a greater role in water resource and environmental conservation in their own communities, with support from the public sector, academic experts, NGOs, and business enterprises.

The first target is to ensure water quality does not fall below 1996 standards in rivers, seas, coastal areas, and all natural water resources, with particular emphasis on the lower Chao Phraya river, the Tha Chin river, pollution control zones, and major tourist destinations. This will ensure conditions are appropriate to sustain aquatic life.

The second target is to formulate a plan for the rehabilitation of Thailand's marine environments. This plan will focus on the conservation, rehabilitation, and proper utilization of natural marine resources and environments, particularly water quality, marine fauna, coral reefs, sea grass, and coastal areas.

Strategies for water resource management

The Eighth Plan proposes the following major strategies to achieve the objectives and targets set for natural resource and environmental management:

- rehabilitation of water resources
- promotion of the participation of local people and communities
- proper management of water resources

The Eighth Plan proposes a number of development guidelines for the rehabilitation of water resources in order to promote balance in the ecosystem and upgrade the quality of life for Thai people, so that they can contribute towards sustainable national development.

- Promote the conservation of land and water resources, including the improvement of soil quality by organic methods. Emphasis should be placed on the promotion of accepted and transferable farming practices, such as integrated farming to replace monoculture, shifting from chemical to organic fertilizers, and terrace farming.
- Reduce the volume and distribution of pollution in local environments by proper management of various types of pollution, such as community and industrial wastewater, air pollution, industrial waste, and hazardous substances, so they do not pose a threat to public health and living conditions.
- Formulate pollution control plans for 25 major river basins around the country.
- Designate guidelines and emergency operational plans to prevent the spread of pollution which affects the quality of terrestrial water resources and marine waters.

Institutional framework

Many government departments are involved in water resources development and water pollution management and control. At present, there are 31 agencies and 17 committees under eight ministries responsible for various aspects of the administration of water resources development (NESDB 1997). Each agency had its own allocated budget and programme to develop water resources for its own specific purposes. The agencies at the departmental level involved in water resources development are listed in Table 6.7.

Sources of information

Information on water resources development, water pollution control, and the management plan is essential to decision-making processes and in the formulation of the master plan for environment and natural resources management. Major sources of information are the related agencies and ministries. Data on water quality from 14 automatic monitoring stations in Thailand are available at the Pollution Control Department (PCD) through its website at www.pcd.go.th. The ERTC conducts different research studies on toxic contaminants in different water resources and provides training related to water quality monitoring, analysis, and management. The Office of Environmental Policy and Planning also prepares the *State of Environment Report of Thailand* annually; information related to this report is now available on a website at www.oepp.go.th. The

Table 6.7 Government agencies involved in water-related government issues

Ministry of Agriculture and Cooperatives	**Ministry of Health**
Office of the Permanent Secretary	Department of Public Health
Royal Rain Making Research and Development Institute	Provincial Health Office
Department of Agriculture	**Office of the Prime Minister**
Royal Forest Department	Electricity Generating Authority of Thailand
Cooperatives Promotion Department	Office of National Socio-Economic Development Board
Department of Fisheries	**Ministry of Industry**
Department of Livestock Development	Department of Mineral Resources
Agricultural Land Reform Office	Department of Industrial Works
Royal Irrigation Department	**Ministry of Communications**
Office of Accelerated Rural Development	Department of Harbours
Department of Community Development	Meteorological Department
Department of Lands	**Ministry of Defence**
Ministry of Science, Technology and Environment	Naval Hydrographic Department
Pollution Control Department	National Security Command Headquarters
Department of Environmental Quality Promotion	**Ministry of Interior**
Office of Environmental Policy and Planning	Local Administration Department
Wastewater Management Authority	Bangkok Metropolitan Administration
	Public Works Department
	Metropolitan Waterworks Authority
	Provincial Waterworks Authority

Source: ERTC (1999)

MWA is at the stage of upgrading its information network to link within and outside of the MWA organization through "Water Net", which is accessible through the MWA website at www.mwa.or.th.

Civil society participation

The recognition of the importance of public participation in the sustainable development of Thailand was furthered by the 1992 Environmental Quality Promotion and Conservation Act. The Act provides for:

- the right of the public in getting access to information;
- the establishment of volunteers in assisting government agencies in environmental matters;
- public participation with the local authorities in pollution control;

- the opportunity for NGOs to register as groups working on natural resource conservation and environmental protection with the Ministry of Science, Technology, and Environment.

The government has set up a guidance framework for public participation through a systematic approach and the provision of appropriate legislation so as to achieve effective management of water resources. There are a number of key action items under the framework.

- Organizing supervisory and coordinating mechanisms for the development of water resources at both national and river basin levels, so as to ensure consistency and continuity in the work of all related agencies.
- With the participation of all parties concerned, setting up appropriate systems at various levels for the allocation of water resources between the various types of water consumer, based on the principles of necessity, priority, and fairness.
- Collecting fees for raw water from industrial and agricultural producers and domestic consumers. The pricing structure for domestic consumption and industrial production will be adjusted to reflect properly the actual costs of procurement, production, distribution, and wastewater treatment.

Table 6.8 Public participation programme on water resource development and water pollution control

NGO	Activities
Thai Rural Reconstruction Movement	Rural development and environment
Central Committee on Farmers' Groups in Thailand	Water resource development in rural areas
Thai Environment and Community Development Association (Magic Eyes)	Rural development issues, encompassing water quantity and quality, health, economic, and civil activities
Svita Foundation	Clean up Chao Phraya river with Magic Eyes; Chao Phraya river's awareness programme
Thailand Development Research Institute	National point of contact for information exchange and coordination
Thailand Environment Institute	Undertake research and provide policy advice to the government on environmental issues
Society for the Conservation of National Treasures and Environment	National point of contact for information exchange and coordination

Source: ERTC (1998)

- Improving the transmission and allocation systems for both irrigation and domestic consumption in communities, so as to minimize wastage of clean water through leaks.
- Conducting public information campaigns to promote thrifty and effective use of water, and encouraging the utilization of water-saving devices and the reuse of cooling water and treated wastewater in some industrial activities.

The Department of Environmental Quality Promotion (DEQP) is the registration body for NGOs working in natural resources and environmental conservation. At present about 197 NGOs work on natural resource and environmental development, and 96 of them are registered with the DEQP. The main objective of this programme is supply financial support for development activities from the Environmental Fund of the Ministry of Science, Technology, and Environment; 20 million baht from the Environmental Fund was provided for promoting the public participation programme performed by the NGOs. Currently, about 15 NGO projects are supported by the Environmental Fund. The government further encourages NGOs to apply for the support from the fund. Information related to the public participation programme on water resource development and water pollution control is shown in Table 6.8. NGOs have been actively participating in the formulation of the Eighth National Plan and have organized major national conferences on the environment which involved 31 NGOs in a forum for expressing the concerns of the NGOs about the state of Thailand's environment.

REFERENCES

ERTC. 1998. *Annual Research Report.* Pathumthani: Environmental Research and Training Center.

IEAT, 1999. *Industrial Estate Wastewater Facilities.* Bangkok: Industrial Estate Authority of Thailand.

Metropolitan Waterworks Authority. 1997. *Report on MWA Treatment Plants.* Bangkok: MWA, Water Quality Control Division.

NESDB. 1997. *Summary of the Eighth National Economic and Social Development Plan (1997–2001).* Bangkok: National Economic and Social Development Board, Office of the Prime Minister.

RID. 1995. *Reservoirs and Water Available, Thailand.* Bangkok: Royal Irrigation Department.

Case studies of air pollution

7

Managing air pollution problems in Korea

Meehye Lee and Zafar Adeel

There is unequivocal evidence that the concentrations of many environmentally important chemical species are increasing in the troposphere as a result of human activity (IPCC 1996). These include greenhouse gases such as carbon dioxide (CO_2), methane (CH_4), nitrous oxide (N_2O), and chlorofluorocarbons (CFCs), stratospheric ozone-depleting species such as CFCs and N_2O, acids, aerosols, and species that directly harm biota such as ozone (O_3) and acids. The need to understand these anthropogenic perturbations to tropospheric chemistry better, particularly in the context of global environmental change and human welfare, has generated a wide research effort over the past few years.

In Korea, the quality of air has been degraded due to rapid development and industrialization in the last couple of decades. Photochemical smog and acid deposition are the most significant problems regarding air pollution. In addition, yellow sand transported from the desert areas of China decreases visibility and has other substantial impacts on air quality. During the 1990s – with a growing awareness of the seriousness of air pollution, partly reflected in Table 7.1 – much effort was put into improving air quality.

In Korea, about 85 per cent of the population live in urban areas that cover 15 per cent of the territory. The population density of Korea is among the highest in the world, and most big cities suffer from frequent smog due to heavy traffic (Ghim 1997). There have been relatively extensive studies on urban smog in metropolitan cities of Korea (Chung

Table 7.1 Nationwide complaints in Korea regarding air pollution

	1989	1990	1991	1992	1993
Total	1,201	1,033	1,274	1,153	2,144
Odour	148	137	175	161	274
Air	179	126	267	233	421

1991; Choi *et al.* 1993; Moon *et al.* 1994; Ghim 1997). The findings of these studies and general public awareness have resulted in gradual progress and the annual average concentrations of sulphur dioxide (SO_2) and total suspended particulate (TSP) have been reduced in most cities. With an increase in traffic levels, however, concentrations of O_3 in major cities have slightly increased. Of the air pollutants for which air quality standards have been established, ozone remains the most resistant to efforts at abatement (MOE 1998a). This chapter examines the current status of the air pollution, its causes and effects, and the environmental performance of Korea to improve its air quality.

Major classes of air pollutants

Air pollutants are gaseous or particulate matter that have an adverse impact on human health and natural ecosystems. In their gaseous phase, along with SO_2, O_3, and its precursors such as the family of nitrogen oxides (generally represented as NO_x), carbon monoxide (CO) and hydrocarbons have been recognized as key species degrading air quality (Finlayson-Pitts and Pitts 1986). Particles of a size less than 2 μm diameter are largely responsible for worsening visibility in urban areas. Such particles can also enter and irritate respiratory organs through inhalation. Particles over several μm in size are composed of crystal minerals or sea salts which naturally exist. In this chapter, due attention is given to the air pollutants that are particularly important in the Korean context.

Ozone

Ozone is of major concern as a public health hazard and phytotoxicant. It primarily originates from the stratosphere by downward transport and is also produced within the troposphere through photochemical reaction involving CO, volatile organic compounds (VOCs), and NO_x (Logan *et al.* 1981). Ozone is the prime culprit in air pollution, and the regulation to reduce ozone concentration has been gradually strengthened.

Nitrogen oxide species (NO_x)

In the troposphere, NO and NO_2 are in rapid photochemical equilibrium in the daytime, thus it is convenient to refer to these two species as NO_x, the sum of the two (*ibid.*). In the presence of hydrocarbons, NO_x is rapidly converted to organic nitrate by photochemical reactions. The organic nitrate, such as peroxyacetyl nitrate (PAN), provides a temporary reservoir for NO_x, eventually decomposing to supply NO_x far from the source. The other major sink of NO_x is oxidation to nitric acid, which is mostly deposited as acid rain. The concentrations of nitric acid measured on Mount Kwan-Ak were several hundreds parts per trillion by volume (pptv), which was much lower than expected from theoretical calculation (Lee, Kim, and Lee 1996). This study indicates that the removal of nitric acid from the atmosphere by deposition was extremely efficient when urban aerosol concentrations were high.

Non-methane hydrocarbons

The most pronounced feature that distinguishes the chemistry of the urban atmosphere from that of the background atmosphere is the relatively high concentrations of a large number of anthropogenic hydrocarbons. Non-methane hydrocarbons are emitted to the atmosphere by a variety of biogenic and anthropogenic sources. The most abundant biogenic hydrocarbons are isoprene and pinene, which are the major sources of hydrocarbons in rural area. In urban areas, most hydrocarbons are anthropogenically emitted to the atmosphere. The most abundant compounds are isopentane, n-butane, toluene, propane, ethane, n-pantane, ethylene, xylene, and benzene, of which concentrations are over 10s of ppbC (Finlayson-Pitts and Pitts 1986). A large number of polycyclic aromatic hydrocarbons (PAH) are emitted from motor vehicle sources. Some aromatics and PAHs are recognized to have carcinogenic effects (Lee *et al.* 1997).

Sulphur dioxide (SO_2)

Fossil fuel combustion and smelters are major anthropogenic sources of SO_2 released to the atmosphere. The air pollution caused by SO_2 is notorious as London smog – where in 1952 several thousands of people died of smog-related damage (Finlayson-Pitts and Pitts 1986). In Korea, the concentrations of SO_2 have been gradually decreased by changes to energy and abatement policies.

Urban aerosols

The urban aerosol represents a mixture of primary emission particles such as soot and ash – these are typically reported as TSP. Primary particles undergo chemical and physical processes, continually changing chemical compositions and particle size. Aerosol is also generated through gas-to-particle conversion processes. Urban aerosol has a smaller size than that of natural aerosol such as sea salt and mineral dust. Sulphate is the most predominant component of urban aerosol and carbon accounts for about 40 per cent of the total fine particle mass (Seinfeld 1988). The ratio of organic carbon to elemental carbon in fine particles is suggested to be a good indicator of the formation of secondary organic aerosols (Lee *et al.* 1997).

Yellow sand

The advent of yellow sand is a special phenomenon observed over the north-western Pacific regions (Iwasaka *et al.* 1988; Murayama 1988). Yellow sand itself is not an anthropogenic air pollutant. When it is apparent, especially in the early spring, however, it substantially decreases visibility and can cause respiratory diseases. The reported reduction in visibility due to yellow sand has been up to one kilometre (MOE 1994). In January 1999 yellow sand was observed all over Korea – which was the first event of its kind during the winter season. In Korea, various approaches to yellow sand research have been made, including field measurement, theoretical model studies, and remote sensing, through the National Institute of Environmental Research, the Meteorological Research Institute, and universities (MOST 1989–1991; MOE 1994).

Acids

Sulphate and nitrate are the most important compounds generating acidity in the atmosphere. In addition to these compounds, small amounts of organic acids such as formic acid and acetic acid also contribute to acidity. Due to the presence of these acids, the pH of fog, cloud droplets, and rain decreases to below 5.6. Damage due to acid rain has been reported all over the world, and it is known that the effect is detrimental to plankton and fish in lakes and coastal regions and forests located close to big cities.

Offensive odours

A range of chemical compounds give off offensive odours, including sulphide, mercaptan, ammonia, aldehydes, and aromatics. There are a variety of sources of odours and differences in personal sensitivity to them. Industrial odours are generated by production facilities, and offensive odours are also generated in the processes of microbial degradation of organic wastes. In addition, the degree of pollution due to offensive odour is largely dependent on meteorological conditions. Hence, it is complicated and subtle to define and solve the problems arising from offensive odours (Kim 1993).

Status of air pollution in Korea

Sources and emission

Air pollutants generally come from both natural sources such as volcanic emissions or mineral dust and anthropogenic sources. Anthropogenic sources of air pollutants are further classified into point source, area source, and line or mobile source. Point source refers to a source of emission at a specified place, such as power generators, waste incinerators, and other stationary industrial emissions. Area source is some specified region with a number of facilities generating air pollutants, such as compact residential apartment complexes. Automobiles are dominant mobile sources of pollutants affecting areas along the roads.

According to the statistics of the Ministry of Environment, transportation accounts for 48.8 per cent of all pollutants emitted to the atmosphere (MOE 1995). The contribution made by industry is 28.8 per cent and the rest is emitted from the public utility sector: power generation – 14.6 per cent and heating – 8.8 per cent. Total emissions of CO_2, SO_2, CO, hydrocarbons, NO_2, CFCs, and TSP are shown for 1991–1997 in Figure 7.1. Figure 7.2 shows the contribution of mobile sources, power generation, industrial sources, and heating to total emissions of SO_2, NO_2, CO, hydrocarbons, and TSP during 1997.

Emissions of SO_2 have been steadily decreasing in Korea since 1990. This can be attributed to the use of fuel with low sulphur content and LNG. SO_2 emissions are, however, still between two and five times higher than those of developed countries except the USA and UK, while total energy consumption is about one-third to two-thirds of the level of consumption in those countries. A theoretical study has predicted that in 2010 the emissions of SO_2 from the industrial sector would be 2.8 times those of 1993 and the concentrations of SO_2 would exceed the standard in about one-third of the cities in Korea (Kim 1997). CO emissions have been substantially reduced over the period studied in Figure 7.1. Any substantial change in the emissions of TSP and hydrocarbons is not noticeable during this period. However, the emissions of NO_2 continue to

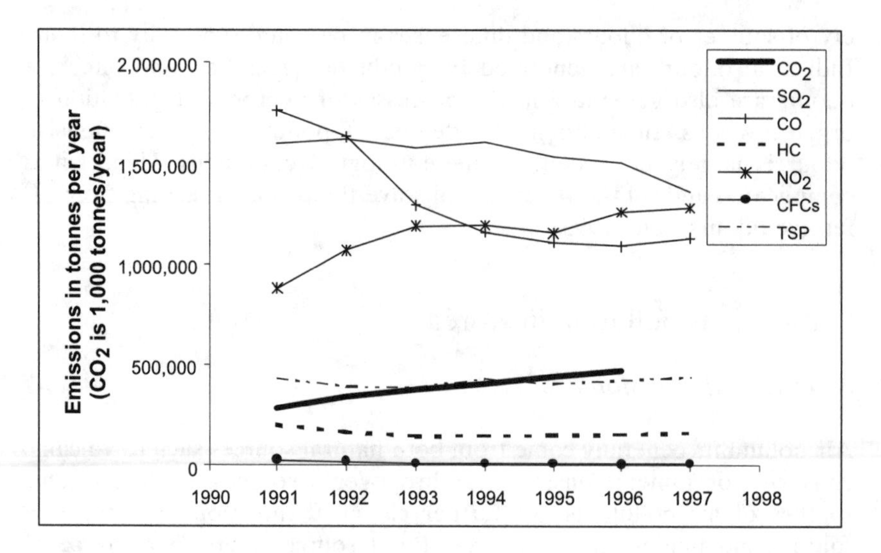

Figure 7.1 Annual emissions of SO_2, CO, hydrocarbons, CO_2, NO_2, CFCs, and TSP, 1991–1997
Source: MOE (1998a)

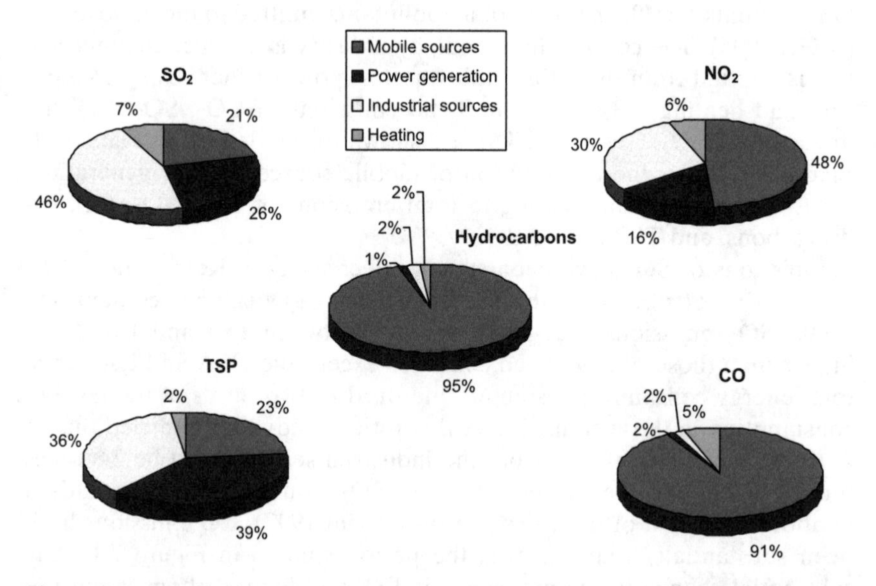

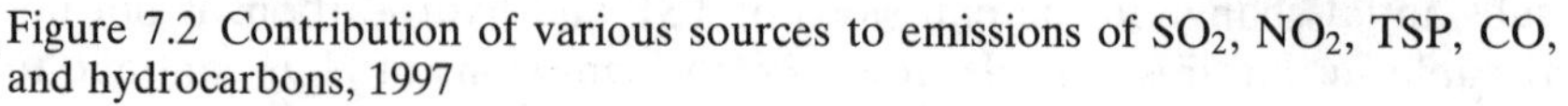
Figure 7.2 Contribution of various sources to emissions of SO_2, NO_2, TSP, CO, and hydrocarbons, 1997

increase. Annual NO_2 emissions are twice as much as those of Japan but about one-third of those of the USA.

It can be observed from Figure 7.2 that industrial emissions were the major source of SO_2 (about 46 per cent) and TSP (about 38 per cent) during 1997; this trend stayed more or less the same over the period 1990–1997. The emission of SO_2 from heating decreased remarkably, while the contribution from mobile sources increased over the same eight-year period. Apportionment of emission sources for SO_2 is similar to that of TSP. For TSP, power generation is also an important source. In 1991, CO emissions from transportation and heating were balanced. Over the next six years, the proportion of transportation rapidly increased and the contributions from other sources were insignificant. Likewise, transportation is the most predominant source of hydrocarbons. For NO_2, mobile sources account for about half of the total emissions and industrial sources account for about 30 per cent. These results indicate that air pollution by industries and other public utilities has been brought under control to some extent, but the same cannot be said of the transportation sector. Moreover, NO_2 and hydrocarbon emissions from mobile sources have increased, leading to an increase in ozone concentrations in metropolitan cities.

In 1965 there were only 40,000 automobiles in the nation, but the number exponentially increased over the next three decades and exceeded 10 million in 1997 (MOE 1998a). Gasoline, diesel, and LPG vehicles account for 68.2 per cent, 29.4 per cent, and 2.5 per cent of the total number of vehicles, respectively. Amongst the different types of automobiles, passenger cars have increased most dramatically, multiplying their total number by more than 400 times. The overall total number of motor vehicles was expected to reach 14 million by the year 2000. About the half of these vehicles are operated in the six largest cities of Korea (MOE 1995).

The ratio of automobile emissions to total emissions is over 70 per cent in Seoul, Taejon, and Kwangju, which are populated cities with fewer industrial complexes compared to Inchon or Pusan (see Table 7.2). It is therefore evident that air pollution due to automobile emissions is an

Table 7.2 Automobile emissions in Korea's six largest cities (in thousands of tonnes)

	Seoul	Pusan	Inchon	Taegu	Taejon	Kwangju
Automobile emission	351	139	83	84	46	46
Total emission	455	434	250	129	63	62
Percentage	77	32	33	65	74	74

imminent threat to metropolitan areas. As is apparent from Figure 7.2, motor vehicles are also the most significant source of hydrocarbons. Most hydrocarbons emitted to the atmosphere are volatile organic compounds (VOCs) and motor vehicles account for about 50 per cent of the total VOC emissions. Interestingly, VOC emissions are also contributed by sources that are typically not governed by air pollution control laws – these include painting facilities, petrol stations, asphalt manufacturing and application, and dry-cleaning services.

Overall state of air quality in Korea

Figure 7.3 shows the annual average concentrations for selected species measured in Seoul and Pusan between 1990 and 1997 (MOE 1998a).

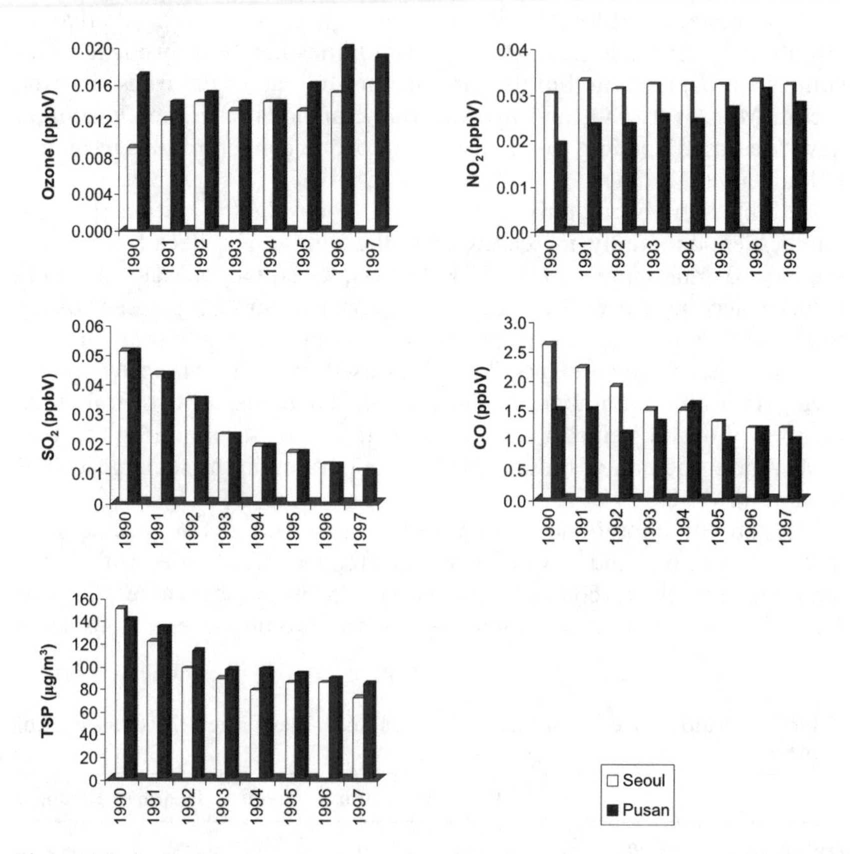

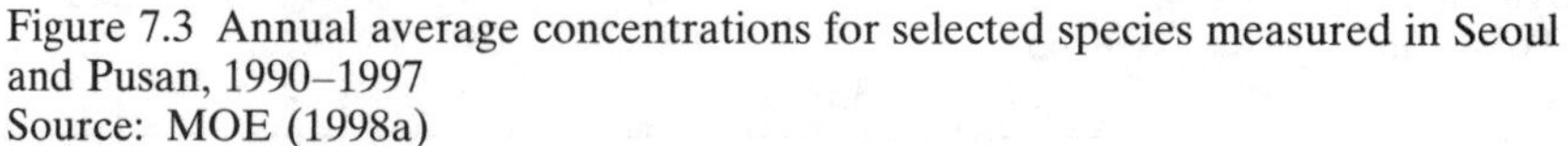
Figure 7.3 Annual average concentrations for selected species measured in Seoul and Pusan, 1990–1997
Source: MOE (1998a)

Seoul is selected as a representative of a more urbanized environment, whereas Pusan represents a greater presence of industrial units in an urban setting.

Ozone concentrations have been increasing since 1989 and are expected to rise further (Figure 7.3). This is, in part, a direct result of increased motor vehicle emissions – a trend which is similar for both Seoul and Pusan. Photochemical smog is the most serious form of air pollution in metropolitan areas, where the number of days of maximum O_3 concentration over 100 ppb is increasing (Choi 1993; Moon 1994; Ghim 1997). Since O_3 is a secondary pollutant produced by photochemical reactions, its concentration varies diurnally and when smog was observed over a whole day the maximum value was often found to exceed the minimum value by well over 10 times, especially in the summer.

Concentrations of NO_2 measured in these cities are around 300 ppt and have slightly increased during the last decade. In Seoul, there is not much change in NO_2 concentrations during the period presented in Figure 7.3 and concentrations are higher than Pusan, Taegu, and Inchon. Like O_3 the level of NO_2 is expected to increase due to an increase in car emissions. Because of the difficulties in reducing NO_x emissions in conventional combustion processes, the need for development of new and efficient technologies to reduce NO_x concentrations is highlighted.

SO_2 concentrations have declined considerably and consistently during the 1990s, which is mostly due to the use of clean fuel containing less sulphur. In cities with large chemical industry complexes such as Yochon, Ulsan, and Seosan, high emissions of SO_2 could have a serious impact on the local environment, even though annual average concentrations are reduced (Kim and Lee 1997). Decrease in CO and TSP concentrations is apparent in Figure 7.3, which also reflects a nationwide trend. Acidity of rainwater has been measured in 30 cities since 1983. However, there was no clear change in pH values during the last decade – albeit there was some annual fluctuation.

Yellow sand originating in China frequently reduces the air quality in Korea. TSP often exceeds the standard when yellow sand is transported across the Korean peninsula, particularly at low altitudes. The total amount of yellow sand produced in China is estimated as 20 Tg/yr. A recent numerical model simulation predicts the annual flux of yellow sand to the Korean peninsula to be 2.7–8.9 Tg, with dry deposition of 2.1–490 Gg/km^2 and wet deposition of 1.5–65 Gg/km^2 (Chung and Park 1998). The abundance of heavy metals on the surface of the mineral dust is also a big public concern regarding yellow sand. The frequency of yellow sand incidents is presented in Table 7.3.

There are no data available concerning measurements of individual hydrocarbons in urban areas. Na and colleagues (Na, Kim, and Ghim

Table 7.3 The frequency of yellow sand incidents observed in Korea

Year	1983	1984	1985	1986	1987	1988	1989	1990	1991	1992	1993	1994
Number	5	5	2	–	1	8	–	3	10	8	14	–

1998; Na *et al.* 1998) measured volatile organic compounds in Seoul and Ulsan. In Seoul, propane, ethylene, and toluene were the most abundant species, which indicates that car exhaust, LPG, and gasoline evaporation were the main sources of VOCs observed in the Seoul metropolitan area. In Ulsan, ethylene was the predominant compound over both the downtown area and industrial complexes. VOC concentrations measured at industrial sites were about twice as high as those in the downtown area. In Seoul, VOC concentrations were higher in the winter than in the summer – this is due to the longer lifetime of VOCs compared to photochemical oxidation and increased emission of VOCs during winter.

Regulatory framework in Korea

Environmental administration

The Ministry of Environment (MOE) was established as a Cabinet-level ministry in 1990 and its function has been further strengthened since. The MOE is responsible for making policies and devising comprehensive plans to prevent air pollution and preserve the natural environment. The National Institute of Environmental Research (NIER) supports the MOE through monitoring, research, tests, and assessments related to air pollution. The NIER was separated from the National Health Research Institute in 1978 as a specialized research institute on the environment and was transferred to the environment administration in 1980. For the education and training of government officials, technicians, and environmental managers, the Environmental Officials Training Institute was founded in 1990 under the NIER. To settle disputes over environmental pollution, the Central Environmental Dispute Settlement Commission is located at headquarters and Regional Environmental Dispute Settlement Commissions are located in Seoul and in the six largest cities and provinces. To achieve effective environmental management that takes full account of local conditions, environmental management offices were set up in six regions under the MOE. Their duties are monitoring and regulation of air quality, enforcement, and environmental impact assessment of the air.

Related ministries

Ministries other than the MOE are also directly or indirectly responsible for various aspects of managing air pollution. The names of the ministries and their respective involvements are listed below.

- Ministry of Science and Technology – regulation of transportation, handling, disposal, and treatment of nuclear and radioactive industrial wastes.
- Ministry of Agriculture and Forestry – measurement of agriculture and forestry pollution.
- Ministry of Trade, Industry, and Energy – import and export of toxic substances and import restriction on industrial wastes; allocation and management of industrial sites; supply of low-sulphur oil; research and development on new and alternative energy sources.
- Ministry of Construction and Transportation – approval and performance testing of motor vehicles.
- Ministry of Labour – countermeasures against occupational diseases and improvement of working conditions.
- Office of Forestry – protection of forests and monitoring of forest destruction activities.

Korean national environmental legislation

The Basic Environmental Policy Act is the basis of all environmental laws. With the objective of preserving the nation's environmental resources, the Act provides basic policies for environmental protection and clarifies the main objectives and directions of these policies. More than 20 environmental laws have been passed by the National Assembly. Of these, selected laws related to air quality preservation are given in Table 7.4.

The formation and development of environmental laws and regulations are closely related to the public perception of environmental problems. In Korea, environmental problems emerged in the 1960s when the country adopted growth-oriented economic policies with the launch of the first five-year economic development plan. To address environmental problems arising from industrialization, the Pollution Prevention Act was enacted in 1963, which is regarded as the first environmental law in Korea. However, the law was never used effectively to deal with air pollution problems. The Act was extensively revised in 1971 to introduce permissible emission standards.

As a result of rapid economic development, air pollution spread pervasively over the country and public concerns about environmental pollution grew. In response, the Pollution Prevention Act was replaced with the Environmental Conservation Act in 1977. In 1980, environmental

Table 7.4 Legislation related to air pollution

Act	Year issued	Year amended	Goal and content
Environmental Conservation Act	1977	1986	Establish environmental impact assessment procedure and environmental standards for air as well as water, noise, solid waste, etc.
Environmental Management Corporation Act	1983	1993	Found Environmental Management Corporation for the efficient and effective management of programmes regarding pollution control
Basic Environmental Policy Act	1990		Prevent danger from pollution and preserve natural environment Define duties of citizens and government
Toxic Chemical Control Act	1990	1991	Establish comprehensive management system and strengthen safety inspection for chemical products
Environmental Dispute Settlement Act	1990		Provide procedures for mediation, reconciliation, and adjustment of disputes due to pollution
Air Quality Preservation Act	1990	1995	Prevent damage by air pollution Establish air quality monitoring network and set emission control standards
Environmental Impact Assessment Act	1993		Designate the scope of coverage, items, and standards for assessment
Act for the Support and Development of Environmental Technology	1994		Develop and promote domestic technology which is environmentally sound

administration was separated from the Ministry of Health and Social Affairs to cope effectively with environmental problems, and environmental rights were added to the constitution as a basic human right. However, air pollution worsened and public demand for clean air increased during the 1980s. Thus the government tightened environmental standards and performed environmental impact assessments.

In 1990, the Environmental Preservation Act was further divided into four laws: the Air Quality Preservation Act, the Water Quality Preservation Act, the Noise and Vibration Control Act, and the Toxic Chemical Control Act. These four laws were supplemented and strengthened with two new Acts: the Basic Environmental Policy Act and Environmental Dispute Settlement Act. To prevent damage to human health and the environment due to air pollution, the Air Quality Preservation Act sets permissible emission standards for specific air pollutants and imposes fines for pollutant discharge.

The Korean national government also introduced a national action plan to protect the atmosphere. The purpose, in part, was to reduce emissions of air pollutants and improve air quality. Specific objectives are as follows:

- improvement of the emission charges system, urban transportation systems, and environmental standards
- implementation of sustainable energy policy and appropriate treatment policy for volatile organic compounds
- reduction of automobile emissions and development of fossil fuel substitutes
- introduction of ozone warning system
- research on transboundary air pollution.

Of these, the ozone warning system was first introduced in Seoul in 1995 and was extended to large cities across the nation from 1997. It will be developed as an air pollution forecasting system to predict the conditions for serious air pollution, including high ozone concentrations, and minimize irreversible damage due to air pollution. It is already running in metropolitan cities such as Tokyo and Osaka in Japan, and Los Angeles in the USA. As shown in Table 7.5, ozone warning is divided into three steps by the level of ozone concentration.

Table 7.5 Ozone alert system in Korea

Warning type	Start	End
Warning	≥0.12 ppm	<0.12 ppm
Alert	≥0.3 ppm	<0.3 ppm
Emergency	≥0.5 ppm	<0.5 ppm

Table 7.6 Proposed environmental standards for selected air pollutants

Species	Present standard	Improvement goal	WHO standard
SO_2 (ppm, annual average)	0.03	0.02	0.015–0.023
TSP (μg/m^3, annual average)	150	100	60–90
PM-10 (μg/m^3, annual average)	150	100	70

The Korean government also set up the Second Mid-term Comprehensive Plan for Environmental Improvement (1997–2001) in 1997. The main purpose was to conserve and maintain the national territory as a healthy and comfortable living environment so that human beings and nature harmoniously coexist in it and it can be passed down through generations to come. As part of this plan, the government will strengthen some of the standards for air pollutants such as SO_2 that have not met the WHO's recommended level, and will have local governments set and operate their own regional environmental standards within national limits. Some selected standards to be revised are presented in Table 7.6. Additionally, because automobiles are a major source of urban air pollution, the Korean government plans to strengthen vehicle emission standards to the level of advanced nations, develop and distribute natural gas and electric vehicles, and extend toll collection on congested roads. For example, the SO_2 emissions standard will be lowered from 0.013 ppm to 0.01 ppm in Seoul and the total vehicle-based emissions will be reduced from 2,050,000 tonnes to 970,000 tonnes.

Declaring the twenty-first century as "the century of the environment", Green Vision 21 was approved by the government in 1995 with the objective of achieving environmentally sound and sustainable development. It provides a set of policies and some quantitative targets to be achieved by 2005. Specific goals are to expand clean fuel supply; to reduce automobile pollution; and to create a pleasant air quality in underground spaces including subway stations.

Emissions standards

To improve air quality, emission standards are necessary for each sector generating air pollutants, which is the first step for air pollution control. The Air Quality Preservation Act prescribes emission limits for 26 substances including O_3, CO, SO_2, NO_x, and TSP. These limits were tightened by about 80 per cent in 1991, as shown in Table 7.7, and were still being strengthened in 1999. There are two types of regulatory standards: the first one is generally applied to all regions of the country; the second

Table 7.7 Emission standards for some selected air pollutants

Pollutants	Source	Emission standards	
		Before 1 January 1991	Since 1 January 1991
CO	Power plant or boiler		
	Liquid fuel	≤350 ppm	≤350 ppm
	Solid fuel	≤400 ppm	≤400 ppm
	Incinerator	≤600 ppm	≤600 ppm
	Cement production	≤600 ppm	≤600 ppm
	Others	≤700 ppm	≤700 ppm
HCl	HCl production	≤15 ppm	≤6 ppm
	Phosphate production	≤2 ppm	≤0.6 ppm
	Chemical fertilizer production	≤10 ppm	≤10 ppm
	Acid treatment of metal surfaces	≤5 ppm	≤2 ppm
	Incinerator	≤60 ppm	≤50 ppm
	Glass production	≤2 ppm	≤0.6 ppm
	Others	≤6 ppm	≤6 ppm
Chloride	Incinerator	≤60 ppm	≤60 ppm
	Others	≤10 ppm	≤10 ppm
SO_2	**Boiler**		
	Liquid fuel		
	Low sulphur content		
	≤1.0%	≤540 ppm	≤540 ppm
	≤0.5%	≤270 ppm	≤270 ppm
	≤0.3%	–	≤180 ppm
	Others	≤1,950 ppm	≤540 ppm
	Solid fuel		
	Solid fuel restriction zone	≤250 ppm	≤250 ppm
	Others		
	Smokeless coal produced in Korea	≤700 ppm	≤500 ppm
	Other solid fuel	≤500 ppm	≤250 ppm
	Power plant		
	(New construction built after 1997)		
	Liquid fuel	≤120 ppm	≤120 ppm
	Solid fuel	≤120 ppm	≤120 ppm
	Petroleum coke production	≤700 ppm	≤270 ppm
	Other solid fuel	≤500 ppm	≤270 ppm
	Metal refining	≤650 ppm	≤650 ppm
	Sulphuric acid production		
	Sulphur burning		
	Other processes	≤300 ppm	≤300 ppm
	Chemical fertilizer production	≤200 ppm	≤200 ppm
	Petroleum refining process	≤350 ppm	≤350 ppm
	Desulphurization	≤500 ppm	≤300 ppm
	Coke production	≤150 ppm	≤150 ppm
	Incinerator	≤300 ppm	≤300 ppm
	Others	≤500 ppm	≤500 ppm

Table 7.7 (cont.)

		Emission standards	
Pollutants	Source	Before 1 January 1991	Since 1 January 1991
NO_2	**Liquid fuel**		
	Power plant	≤400 ppm	≤950 ppm
	Other facilities	≤250 ppm	≤250 ppm
	Solid fuel	≤350 ppm	≤350 ppm
	Gas fuel		
	Power plant	≤500 ppm	≤500 ppm
	Other facilities	≤400 ppm	≤400 ppm
	Others	≤200 ppm	≤200 ppm
CH_2O	All	≤20 ppm	≤20 ppm
Benzene	All	≤50 ppm	≤50 ppm
Phenol	All	≤10 ppm	≤10 ppm

one is a set of stricter standards that are applied, as exceptions, to special regions designated as air preservation zones. In the latter case, the standards are applied because general standards would hardly have any impact on the air pollution situation. The general emission standards for some selected pollutants are given in Table 7.7.

NO_2 emissions limits varied according to the type of facilities, and were in the range of 200–1,400 ppm until 1988, then lowered to the range of 200–950 ppm by 1999. CO emission standards for passenger cars made after 1988 were lowered to 1.2 per cent from 4.5 per cent, although remaining at 4.5 per cent for other types of automobiles. For VOC emissions, the ozone-producing potency or carcinogenic effect is different from species to species. Therefore, it is critical to estimate total emissions and establish emission limits for individual VOCs. To accomplish this, reliable measurements must first be made in Korea.

Air quality standards

Air quality standards are targets for air pollution control. To meet these standards, emission standards are modified or tightened. Air quality standards for SO_2, CO, NO_2, particulate, O_3, and Pb are listed in Table 7.8.

These standards can be compared with those developed by the European Community (EC). For example, the EC applies an annual average SO_2 standard of 0.03 ppm when TSP concentration is also high; but a

Table 7.8 Regulatory standards for air pollutants

Pollutants	Standard
SO_2	≤0.03 ppm : annual average ≤0.14 ppm : 1 day average ≤0.25 ppm : 1 hour average
CO	≤9 ppm : 8 hour average ≤25 ppm : 1 hour average
NO_2	≤0.05 ppm : annual average ≤0.08 ppm : 1 day average ≤0.15 ppm : 1 hour average
Particulate	
Total	≤150 μg/m^3 : annual average ≤300 μg/m^3 : 1 day average
PM-10	≤80 μg/m^3 : annual average ≤150 μg/m^3 : 1 day average
O_3	≤0.06 ppm : 8 hour average ≤0.1 ppm : 1 hour average
Pb	≤1.5 μg/m^3 : 3 month average

more stringent standard may be applied in the winter when fuel consumption increases. This flexibility helps to prevent smog formation in urban areas during the winter. For NO_2, the one-hour standard specified in Korea is stricter than that of the WHO (0.21 ppm), while the annual average NO_2 standard is similar to those of other countries.

Economic tools for air pollution management

An emission charge system was put into effect in 1983 to prevent environmental damage due to discharges of pollutants in excess of emission limits to the atmosphere. Companies violating the law must pay fines equivalent to the treatment expenses for the excess volume of pollutant discharged. Emission fines were established for 10 specific air pollutants, and about 7 billion won was collected in 1995 (US$1 = roughly 1,200 won).

Collecting tolls on congested roads, bridges, and tunnels has also been proven as an effective method to control traffic and reduce car emissions caused by congestion. Additionally, gasoline tax is relatively high in Korea to discourage the use of private cars. On the advice of the International Monetary Fund, gasoline tax has been raised even more. The income from this taxation and pricing system is supposed to be used for improving the public transportation system; however, the efficiency of such expenditure has not been clearly demonstrated to date.

Table 7.9 Estimated pollution abatement and control expenditure (billion won)

Sector	1992	1993	1994	1995	(%)
Government	2,232	2,147	2,592	2,915	(49.2)
Industry	2,062	2,042	2,345	2,591	(43.8)
Household	313	372	394	414	(7.0)
Air	840	834	967	1,080	(18.2)
Total	4,607	4,831	5,331	5,920	

Source: OECD (1997)

In Table 7.9, the estimated costs for pollution abatement and control are given for 1992–1995. The total expenditure was substantially raised during that period. Government and industry spent almost the same amount on pollution control. Of the total cost, about one-fifth was spent to reduce emissions to the atmosphere. The government's budget for environmental purposes has also steadily increased. In the Ministry of Environment, 8.9 billion won of a total of 1,080 billion won was used to preserve air quality during 1997 (MOE 1998b). Although the total budget for the MOE continues to increase, the expenditure for air quality preservation in 1997 diminished by about 10 per cent compared to that of 1996. In 1998, even the total expenditure was expected to decrease due to the economic crisis that paralysed economic activity in Korea.

Critical policy issues in air pollution

Precursors of ozone

Ozone concentrations are largely dependent on those of NO_x and volatile organic compounds. Thus controlling ozone concentrations relies upon the emission control of those precursors. The US Environmental Protection Agency's empirical kinetic modelling approach (EKMA) is a useful tool to relate NO_x and hydrocarbon changes to changes in maximum ozone concentrations for particular urban areas (Seinfeld and Pendis 1997). EKMA can be used in determining whether reduction in VOCs or NO_x favours decreasing O_3 under a given set of atmospheric and environmental conditions. The US EPA has encouraged controlling VOCs for ozone reduction because it is cheaper to do so. In spite of the substantial effort and expenditure to control the emission of VOCs and NO_x, the reduction in O_3 has been modest in the USA (*ibid.*).

In Korea, ozone concentrations have continuously increased in the last decade. However, the government has not set an appropriate policy or strategy for ozone abatement. VOCs have recently been considered for regulation, but reliable measurements of VOCs were made only a few years ago (Na, Kim, and Ghim 1998; Na *et al.* 1998). The Air Quality Preservation Act, amended in 1995, stipulates that oil storage facilities and refineries are to be equipped with an emission controller. It will therefore be a critical policy issue to determine to what extent VOC control is invoked for ozone reduction and NO_x controls for further ozone abatement.

Transboundary export of air pollutants

There are several examples available where transboundary fluxes of air pollutants have fuelled an international conflict. For example, acid rain occurring in eastern Canada was caused by the transport of SO_2 from the Midwest of the USA. Thus the USA and Canada began negotiations to limit acid deposition resulting from transport across the border (Schwartz 1988). Arctic haze was due to emissions of pollutants from eastern European countries (Rahn 1981). The damage by biomass burning which was brought about in Indonesia partly due to the 1997–1998 El Niño phenomenon affected neighbouring countries. An "emergency" was declared in Malaysia as well as in Indonesia due to air pollution threatening the lives of people.

Recent studies show that long-range transport of air pollutants from Asia is beginning to have a significant impact on the atmosphere over a large part of the Pacific (Howell *et al.* 1996, 1997). It is more pronounced during the early spring due to the stronger jet air stream. Asian dust is transported from China to Korea by westerly winds and is often contaminated with toxic substances. The "signature" of this dust was found over remote Pacific regions and it significantly affects the Yellow Sea (Gao *et al.* 1992a, 1992b). Korea is located downwind of China, where sulphur emissions account for about 15 per cent of the global total. Although data are still scarce, it is believed that anthropogenic emissions from China play a part in acid deposition over Korea, and Korea also contributes to acid deposition over Japan (Hong, Kim, and Lee 1997). It is therefore difficult to attribute air pollution such as acid and dust deposition in any region of the Pacific to a single contributing emission source. Extreme effort should be exerted to develop strategies for air pollution abatement in Korea. To understand transboundary air pollutant movement and to establish an effective abatement policy, the government needs to foster not only national research but also international cooperation among the East Asian countries.

Activities of key stakeholders

Environmental NGOs

With an increase in public awareness of environmental issues, the activities of environmental non-government organizations (NGOs) have commanded more attention. NGOs play an important role in educating and informing the public. They also keep up public pressure on air pollution problems by conducting campaigns, surveys, and research on critical issues and policies at the local and national scale. Keeping an eye on the activities of government and parliament, they assess and evaluate policies and projects, and propose alternatives if it is appropriate. There are few restrictions on founding environmental NGOs in Korea, with no specific legislation governing the process. Government not only invites NGO representatives to participate in the policy formulation process, but also provides some limited financial support to selected environmental NGOs.

There are about 300 NGOs carrying out a variety of activities in Korea. Of these, the Korean Federation for Environmental Movement (KFEM) founded in 1993 and Green Korean United (GKU) founded in 1991 are the most active. The former runs 33 local offices nationwide and the latter runs eight. Some religious and faith communities such as the YWCA also serve as environmental NGOs. To preserve the earth's environment is regarded as the responsibility of religious people, and they are infused with a vision for the future. These NGOs have established professional relationships with international NGOs such as Greenpeace and the Worldwide Fund for Nature for information exchange and international cooperation.

The activities of the KFEM and GKU during the period 1993–1998 can be summarized as the following:

- monitoring of air quality and mapping the degree of air pollution in Korea
- investigation of air pollution over the Siwha and Yochon industrial complexes
- investigation of air pollution by incineration in Taejeon
- participating in NGOs' ASEM held in Bangkok in 1997
- organizing a seminar on the air pollution in Siheung-Si
- organizing a symposium on environmental policies regarding the Framework Convention on Climate Change (FCCC)
- organizing a human rights movement to support people imprisoned for their anti-government activities related to environmental issues
- performing experiments on the degree of corrosion of metal due to air pollution

- participating in an environmental NGOs' international conference on the roles and strategies of developing countries in preventing climate change (1998).

Role of the industrial sector

The Korean economy has been developed with the support of the government's export-oriented policies and relatively mild environmental regulations. However, environmental awareness continues to grow and governmental regulations have been gradually tightened. In the early 1990s, some conglomerates and large companies started to set up their own environmental research institutes, internal environmental audit systems, and guidelines and criteria. They put much effort into the prevention and abatement of air pollution through two types of activities. Firstly, by adopting and improving environmentally sound technology, they can save energy and raw materials in production and thereby reduce emissions and waste disposal. Secondly, by improving their environmental management systems, they can assess their environmental performance and pinpoint problems as they arise, mitigating the effects on the atmosphere.

The example of large enterprises is expected to induce a more general movement among small and medium-sized industries, including contractors, which are not capable of adopting advanced technology or may have a lower level of environmental awareness. Financial aid is necessary for companies to develop environmentally friendly technology, assess the environmental impact of their products, and improve their environmental management systems. By the end of 1997, however, the economic crisis had struck all activities in Korea. Bankruptcy and unemployment rates were at their highest during 1998. Budget cuts and structural reform in every section of the Korean economy were demanded by the International Monetary Fund. In this process, environmental research institutes belonging to conglomerates and medium-sized companies became the first target for major surgery. As a result, the role of industries in environmental preservation seems to have shrunk somewhat in the last couple of years. However, it is anticipated that it will return to its former level as the Korean economy gets back to its normal condition.

Future outlook

Ambient air quality in metropolitan areas of Korea has steadily improved over the last decade. This was achieved through environmental policies encouraging the use of clean fuel with low sulphur and low lead content, and LNG. The Air Quality Preservation Act enacted in 1990 provided the base for effective air quality management. Standards for permissible emissions and air quality were gradually tightened. Emissions of SO_2 and TSP have slightly decreased and those of CO and hydrocarbons have substantially decreased. On the other hand, NO_x emissions have risen, leading to a steady increase in O_3 concentrations. Although concentrations of SO_2 and TSP have decreased and met the national standards, the degradation of air quality is more seriously felt by the public due to an increase in concentrations of O_3 and soot from traffic emissions. This trend of air pollution indicates that Korea is in transition from developing country to developed country, while the ability to cope with air pollution has been slowly improved.

There is still much room for improvement of air quality and implementation of air quality management policies. Ambient air quality standards still do not meet the WHO standards for NO_x, O_3, and PM-10. In industrial cities, concentrations of some air pollutants frequently exceed national standards. There is a growing awareness of the effect of air pollution on the regional and global environment. Thus improving air quality will be a big challenge for Korea and will make an important contribution to the international effort to preserve the environment. In this context, full consideration should be given to the points listed below by all the relevant stakeholders.

For government

- Establish comprehensive national policies for air pollution abatement and air quality management.
- Strengthen the standards for air quality and regulations for pollutant emissions.
- Enforce reasonable economic tools such as tolls, taxation, and fines.
- Improve the ability of local offices to manage local air quality problems.
- Encourage the development and use of environmentally friendly energy sources and automobiles emitting less pollutants.
- Support development of reliable mechanisms for measurement of air pollution.
- Systematically manage environmental data and information while increasing transparency and access to the public.
- Expand the opportunity for the public and NGOs to participate in development and implementation of environmental policies.
- Support the environmental education of the public.
- Participate in international cooperation to preserve the environment.

For industry

- Invest more money for developing new technologies for reducing emissions.
- Take full responsibility for products, from the producing process to final disposal.
- Establish environmental management systems for the whole process.
- Provide environmental education to workers, technicians, and executives.

For NGOs

- Train experts in each specialized area related to air pollution.
- Build professional relationships with government, industrial sectors, and the public.
- Build coalitions with NGOs from other disciplines.
- Provide guidelines on environmental issues for the public.

For the public

- Practise environmentally sound lifestyles through reducing waste, changing consumption patterns, and recycling.
- Participate in the policy-making processes that are open to the public.
- Monitor environmental performance by the government, industries, and NGOs.
- Observe the rules and laws for preserving the atmospheric environment.

REFERENCES

Choi, D. 1993. *A Study on Characteristics and Photochemistry of Urban Air Pollution*. Seoul: National Institute of Environmental Research.

Chung, K. and S. Park. 1998. "A numerical study on the size and depositions of yellow sand events", *Journal of Korean Air Pollution Research Association*, Vol. 14, pp. 191–208.

Chung, Y. 1991. *A Study on Photochemical Smog of Large Cities in Korea*. Seoul: Korean Society of Environmental Science Research.

Finlayson-Pitts, B. J. and J. N. Pitts Jr. 1986. *Atmospheric Chemistry*. New York: John Wiley & Sons.

Gao, Y., R. Arimoto, R. A. Duce, D. S. Lee, and M. Y. Zhou. 1992a. "Input of atmospheric trace elements and mineral matter to the Yellow Sea during the spring of a low-dust year", *Journal of Geophysical Research*, No. 97, pp. 3767–3777.

Gao, Y., R. Arimoto, M. Y. Zhou, J. T. Merrill, and R. A. Duce. 1992b. "Relationships between the dust concentrations over eastern Asia and the remote north Pacific", *Journal of Geophysical Research*, No. 97, pp. 9867–9872.

Ghim, Y.-S. 1997. "Indication of photochemical air pollution in the greater Seoul area, 1990 to 1995", *Journal of Korean Air Pollution Research Association*, Vol. 13, pp. 41–49.

Hong, M.-S., S.-T. Kim, and D.-S. Lee. 1997. "A study on the quantitative analysis of SO_2 dry deposition in northeastern Asia", *Journal of Korean Air Pollution Research Association*, Vol. 13, pp. 231–241.

Howell, J. M., D. D. Davis, S. C. Liu, R. E. Newell, M. Shipham, H. Akimoto, R. J. McNeal, R. J. Bendura, and J. W. Drewry. 1996. "The Pacific Exploratory Mission-West Phase A: September–October 1991", *Journal of Geophysical Research*, Vol. 101, pp. 1641–1653.

Howell, J. M., D. D. Davis, S. C. Liu, R. E. Newell, H. Akimoto, R. J. McNeal, and R. J. Bendura. 1997. "The Pacific Exploratory Mission-West Phase B: February–March 1994", *Journal of Geophysical Research*, Vol. 102, pp. 28223–28239.

Intergovernmental Panel on Climate Change (IPCC). 1996. *Climate Change 1995: The Science of Climate Change*, ed. J. T. Houghton. Cambridge: Cambridge University Press, 1996.

Iwasaka, Y., M. Yamamoto, R. Imasu, and A. Ono. 1988. "Transport of Asian dust (KOSA) particles: Importance of weak KOSA events on the geochemical cycle of soil particles", *Tellus*, No. 40B, pp. 494–503.

Kim, B.-G., J.-S. Cha, J.-S. Han, L.-S. Park, J.-S. Kim, J.-G. Na, D.-I. Choi, J.-Y. Ahn, and C.-G. Kang. 1997. "Aircraft measurement of SO_2 and NO_x over Yellow Sea area", *Journal of Korean Air Pollution Research Association*, Vol. 13, pp. 361–369.

Kim, H. 1993. *Introduction to Air Pollution*. Seoul: Dongwha Science.

Kim, Y.-P. 1997. "A study on estimation and long-term forecasts of SO_2 pollution in each city and country of Korea", *Journal of Korean Air Pollution Research Association*, Vol. 13, pp. 19–29.

Kim, Y.-P., and J.-H. Lee. 1997. "Concentrations of particulate and gaseous ionic and organic species in the ambient air of the Yochon industrial estate", *Journal of Korean Air Pollution Research Association*, Vol. 13, pp. 269–284.

Lee, H.-M., D.-S. Kim, and J.-H. Lee. 1996. "An assessment of the long-term concentration of heavy metals and associated risk in ambient PM-10", *Journal of Korean Air Pollution Research Association*, Vol. 12, pp. 555–566.

Lee, J.-H., Y.-S. Kim, Y.-T. Ryu, and L. S. Yu. 1997. "A study on the health risk assessment on volatile organic compounds in a petrochemical complex", *Journal of Korean Air Pollution Research Association*, Vol. 13, pp. 257–267.

Logan, J. A., M. J. Prather, S. C. Wofsy, and M. B. McElroy. 1981. "Tropospheric chemistry: A global perspective", *Journal of Geophysical Research*, No. 86, pp. 7210–7254.

MOE. 1994. *A Study of the Effect of Long-range Transported Yellow Sand and Air Pollutants on Korean Environment*. Seoul: Ministry of Environment.

MOE. 1995. *Environmental Statistics Year Book*. Seoul: Ministry of Environment.

MOE. 1998a. *Environmental Statistics Year Book*. Seould: Ministry of Environment.

MOE. 1998b. *White Paper on Environmental Monitoring*. Seoul: Ministry of Environment.

Moon, K.-C. 1994. *Study of Smog in Seoul Area III*. Seoul: Korea Institute of Science and Technology.

MOST. 1989–1991. *A Study on Long-range Transport of Air Pollutants and Acid Deposition to Korea*. Seoul: Ministry of Science and Technology.

Murayama, N. 1988. "Dust cloud 'KOSA' from the east Asian dust storms in 1982–1988 observed by the GMS satellite", *Meteorological Satellite Technical Note*, No. 17, pp. 1–8.

Na, K.-S., Y.-P. Kim, and Y.-S. Ghim. 1998. "Concentrations of C2-C9 volatile organic compounds in ambient air in Seoul", *Journal of Korean Air Pollution Research Association*, Vol. 14, pp. 95–105.

Na, K.-S., Y.-P. Kim, H.-C. Jin, and K.-C. Moon. 1998. "Concentrations of water-soluble particulate, gaseous ions and volatile organic compounds in the ambient air of Ulsan", *Journal of Korean Air Pollution Research Association*, Vol. 14, pp. 281–292.

OECD. 1997. *Environmental Data*. Paris: Organization for Economic Cooperation and Development.

Rahn, K. 1981. "Relative importance of North America and Eurasia as sources of Arctic aerosol", *Atmospheric Environment*, Vol. 15, pp. 1447–1456.

Schwartz, S. 1988. "Acid deposition: Unraveling a regional phenomenon", *Science*, No. 243, pp. 753–763.

Seinfeld, J. 1988. "Urban air pollution: State of the science", *Science*, No. 243, pp. 745–752.

Seinfeld, J. and S. N. Pendis. 1997. *Atmospheric Chemistry and Physics: From Air Pollution to Climate Change*. New York: John Wiley & Sons.

8

The Japanese approach to governance of air pollution problems

Makiko Yamauchi

Historical overview

Japan faced severe air pollution problems, especially in urban areas, during the 1960s and 1970s. A variety of measures were undertaken to address and tackle theses problems, and considerable progress has been achieved since then. Following Japan's rapid economic growth and the change of industrial structure that occurred between the post-Second World War period and the oil crisis in 1973, serious pollution by sulphur oxides (SO_x) had appeared around industrial zones. Local residents increasingly suffered from various diseases, such as the Minamata disease (caused by mercury poisoning), Yokkaichi asthma (caused by sulphur dioxide, SO_2), and Itaiitai disease, which is the so-called "dirty four" (*yondai kougai*).

Historically, pollution problems in Japan can be dated back to the Ashio copper mine case in Tochigi prefecture in the 1880s, where crops and fisheries near the copper mines were withered and damaged due to SO_x and acidic dust. Air and water pollution events also occurred in Osaka, Amagasaki, and Kawasaki in the first half of the twentieth century. As a reaction to these serious pollution cases, the main environmental legislative and institutional framework in Japan emerged during the late 1960s and early 1970s (Japan Environmental Council 2000).

At the national level, the Ministry of International Trade and Industry (MITI) and the Ministry of Health and Welfare (MHW) established their own pollution control divisions in the 1960s. The national legal framework was initiated with the Basic Law for Environmental Pollution Control (BLEPC) in 1967. Afterwards, during the period of the Diet (the Japanese National Assembly) seated in 1970 – which was aptly named the Pollution Diet – 14 laws were passed or amended. A key element of these laws was the introduction of a particular variant of the polluter pays principle (PPP); that is, an enterprise must pay a portion of the cost of pollution control public works equal to its contribution to pollution (Abe and Awaji 1995). At the subnational and prefectural levels, local initiatives played an important role in the development and shaping of environmental policy in Japan. The Local Government Law (1947) initiated major decentralization by giving local governments authority to pass their own ordinances and standards. Local governments of large urban areas, such as Tokyo, Osaka, and Yokohama, used this authority to act on pollution problems from the mid-1950s to the early 1960s (Ueta 1994).

As an institution, the Environment Agency of Japan (EAJ) was established in 1971 to consolidate the pollution control duties of 13 different ministries and agencies. The EAJ was charged with implementation of the BLEPC and laws relating with specific environmental media and pollutants. The limited independence exercised by the EAJ in personnel policy is pointed out as a feature of the EAJ (O'Connor 1994). Meanwhile, measures for air and water pollution control and formulation of environmental impact assessment (EIA) procedure were further advanced during the 1980s. In 1993, the Basic Law on Environment was passed as a milestone for new initiatives responding to the concept of sustainable development and global responsibility (Environment Agency of Japan 2000). In 2001, the EAJ evolved into the Ministry of Environment and is represented in the Cabinet.

State of air pollution in Japan

Sulphur dioxide

Sulphur dioxide (SO_2) occurs as a result of combustion of coal and petroleum that contains sulphur. SO_2 causes not only pollution-related diseases, such as Yokkaichi asthma, but also acid deposition. The mass consumption of fossil fuel during the high-growth period worsened the SO_2 air pollution in Japan to a great extent. However, despite the increase in energy consumption and economic growth, SO_x emission levels clearly dropped from 1970 through the implementation of various pollution control measures. These included a regulation of the total amount of SO_x from factories throughout the country, a regulation of the sulphur con-

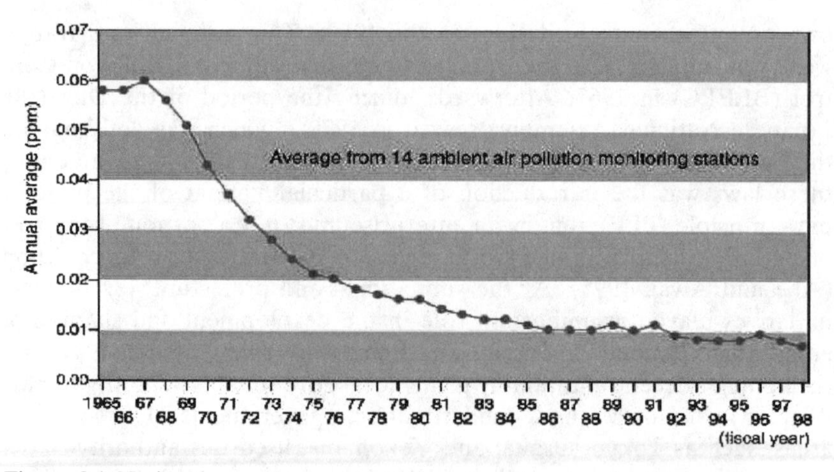

Figure 8.1 National annual average sulphur dioxide concentrations
Source: Environment Agency of Japan (2000)

tent in fuel, and tighter controls for facilities generating soot and smoke. The industry accepted these regulations and implemented additional, expedited positive countermeasures, such as the installation of equipment to desulphurize emitted smoke, the desulphurization of heavy oil, and the import of low-sulphur crude oil. As a result of these changes, the concentration of SO_x in the air had dropped conspicuously from its peak in 1967 (0.059 ppm) by 1998 (0.008 ppm), as shown in Figure 8.1.

Nitrogen oxides

Nitrogen oxides (NO_x), such as nitrogen monoxide (NO) and nitrogen dioxide (NO_2), occur mainly accompanying the combustion of fossil fuels. They come from both stationary sources, such as boilers of factories, and moving sources such as automobiles. NO_x are the causal substances for acid deposition and their high concentration exerts a harmful influence on human health, particularly the respiratory organs. Concentration of NO_x decreased after 1979, but it has started increasing again since 1986. As presented in Figure 8.2, the achievement of the environmental quality standards has been low, particularly in the metropolitan areas of Osaka, Yokohama, and Tokyo.

Carbon monoxide

Carbon monoxide (CO) in the air occurs primarily as a result of the incomplete combustion of automobile fuels. CO influences human health

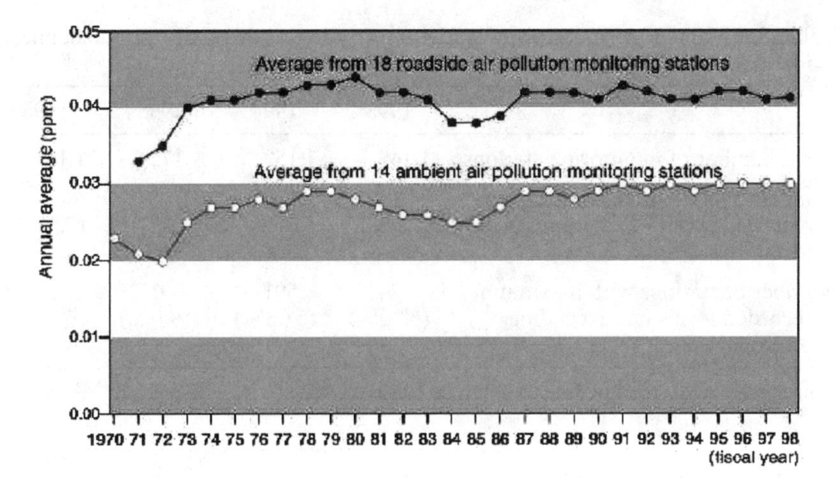

Figure 8.2 National annual average NO_2 concentrations
Source: Environment Agency of Japan (2000)

by uniting with haemoglobin in the blood and hindering the oxygen-carrying function; it also lengthens the life of methane gas, which has a greenhouse effect. An environmental quality standard (EQS) for CO was set in February 1970, and it has been continuously monitored since 1971. All the observation stations in Japan have achieved this EQS, and CO is present at a low level.

Suspended particulate matter

Suspended particulate matters (SPM) are substances found in the form of particles with a diameter of 10 μm or less, such as floating dust, aerosols, etc., that are floating in the air. SPM stays in the air for a long time because of its minute size. It affects the respiratory organs, as it settles in the respiratory system, including the windpipe and lungs. After an EQS was established, measures have been carried out to achieve the EQS standard by regulation of the emission of black smoke from automobiles, and of dust and particles of soot from business and factories. The SPM concentration has remained at almost the same level since 1980. The rate of achievement of the EQS is moving at a low level, and the state of achievement in the Kanto area is especially poor. Further, the fall in the rate of achieving the EQS in the Kyushu, Chugaku, and Shikoku areas is striking. This is assumed to be due to the influence of yellow sand coming across the Sea of Japan, primarily from China.

Table 8.1 Number of monitoring stations recording an excess in photochemical oxidants

	1993	1994	1995	1996
Total number of monitoring stations	1,149	1,159	1,172	1,181
Number of stations with maximum recorded levels of 0.06 ppm or less	12 (1.0%)	5 (0.4%)	8 (0.7%)	3 (0.3%)
Number of stations with maximum recorded levels not exceeding 0.12 ppm	724 (63.0%)	591 (51.0%)	691 (59.0%)	724 (62.8%)

Source: Environment Agency of Japan (2000)

Photochemical oxidants

The level of photochemical oxidants (O_x) measured continuously remains unfavourable in Japan. The recent history of observed levels of photochemical oxidants is shown in Table 8.1. These show the number of stations at which the highest level in one hour during a day (5am to 8pm) was 0.06 ppm or less (the level specified by the environmental quality standard), or less than 0.12 ppm (the level that requires a warning to be issued), respectively. Concentrations of O_x exceeding the 0.12 ppm photochemical oxidants warning standard appeared at 37.2 per cent of monitoring stations in 1996. A warning is issued when the hourly value of O_x concentration is 0.12 ppm or higher and the pollution is considered likely to continue further in view of the meteorological conditions at that time. Such warnings were issued on a total of 95 days in 20 prefectures during 1996.

Compared to other countries in the region, Japan's level of air pollution is relatively low. Of late, its SO_x and NO_x emissions, both per capita and per unit of GDP, have been the lowest in the OECD group. This is largely a result of the initiation and strong implementation of environmental ordinances by local governments. Kitakusyu city, Tokyo prefecture, and Shiga prefecture lead this trend by including provisions for environmental rights and stipulations for planning assessments, setting stricter NO_x standards for automobiles, and promoting citizens' participation. These initiatives complement national legislation. However, ambient air quality in the cities remains a serious problem due to the air pollution caused by motor vehicle traffic. The Ministry of Environment plans to introduce tougher, more comprehensive measures to improve ambient air quality in cities, including tightening controls on vehicle exhaust emissions.

Governance framework for air pollution management

Governance structure

The Environment Ministry is charged with air pollution management as an apex national government agency. Headed by a minister of state who is a member of the Cabinet, it has a primary mandate to execute basic policies and measures for the environment, and to coordinate policies and measures among government ministries and agencies for environmental protection. Since around the 1960s, local governments have established environmental units in order to implement measures delegated by national environmental laws as well as to carry out their own environmental laws and ordinances. Air pollution control is one of the important fields where local government has enforced regulatory, supervisory, and disciplinary measures. The sharing of responsibilities for overall governance of air pollution issues is outlined in Table 8.2 (Abe and Awaji 1995).

Environmental standards are often set at both the national and local levels. Generally, the national government sets the ambient standards, while local governments define specific and effluent standards. Local governments are also responsible for monitoring air quality. Under the Local Government Law, introduced in 1947, they have considerable power in the implementation of their own environmental policy. This law enables municipalities to pass their own ordinances. Since 1970 they have been entitled to set more stringent standards than the national ones. At the same time, by conducting voluntary agreements on pollution control with industry they have been able to achieve more than that envisioned in the national actions and plans.

Legislative framework

The Basic Environment Law

In Japan, the environmental administration was initially implemented under the Basic Law for Environmental Pollution Control (BLEPC) and the Nature Conservation Law. The BLEPC, enacted in 1967, has provided the broad basis for the control of environmental pollution; one of the main elements is the establishment of environmental quality standards to combat the serious industrial pollution which affected Japan badly in the period of rapid economic growth of the late 1950s and 1960s. Prior to the BLEPC, the Soot and Smoke Control Law was enacted in 1962 as the first step in a legislative approach to air pollution. Though this law did help in reducing the levels of dust fall nationwide, it could

Table 8.2 Outline of measures for air pollution control and their jurisdiction

Measures	Jurisdiction
Environmental standards	Ministry of Environment
Emission regulations	
Factories and utilities/business establishments	
Emission standard	Ministry of Environment
More stringent standard	Local government
Structural standard	Environment Agency
Fuel standard	Environment Agency/local government
Monitoring	Local government/Environment Agency
Improvement/implementation	Local government
Power and gas plant	
Improvement/implementation	Ministry of International Trade and Industry (MITI)/local government/ Ministry of Environment
Automobile emissions	
Permissible limits	Ministry of Environment
Safety standard	Ministry of Transport
Monitoring	Local government/Ministry of Environment
Environmental monitoring	Local government/Ministry of Environment
Traffic control	Public Safety Commission at local government/National Police Agency
Plan for pollution control	Ministry of Environment
Research and studies	
Causes and influences	Ministry of Environment
Prevention techniques	MITI
Low-pollution vehicles	MITI/Ministry of Transport
Measures concerning fuel	
Desulphurization	MITI
Nuclear power use	Science and Technology Agency
Working project for pollution control	Ministry of Construction/local government
Financing and assistance	
Financing	Association for Controlling Pollution/ local government
Tax system	Ministry of Finance/Ministry of Home Affairs
Compensation for health damage	Ministry of Environment/local government
Settlement of complaints	Local government
Penalties	Ministry of Justice/National Police Agency

Source: Environment Agency of Japan (1998)

not stop the worsening air pollution from sulphur dioxide because of the development of heavy and chemical industries, the concentration of factories in pre-existing industrial zones, the appearance of new industrial zones, and the rapid growth of urban population. The Nature Conservation Law was enacted in 1972 to bring to an end the destruction of outstanding features of the natural environment.

Although these laws attained considerable results in tackling the environmental problems in that context, conventional methods which focus on regulation became more and more insufficient in dealing with serious new environmental problems. To improve the situation, the Basic Environment Law bill was unanimously passed by the National Diet in its 128th session on 12 November 1993; this law replaced the earlier legislation. The new Basic Environment Law establishes the foundation of Japanese environmental laws and policies and provides basic principles and policy directions to implement a comprehensive environmental policy in this age of globalization.

Under the law, three fundamental principles of environmental policy were established.

- The blessings of the environment should be enjoyed by the present generation and passed on to future generations.
- A sustainable society should be created where environmental burdens caused by human activities are minimized.
- Japan should contribute actively to global environmental conservation through international cooperation.

The law also defines the responsibilities of each party in society – the state, the local governments, the industrial corporations, and the general public. These parties should make cooperative efforts to protect the environment through fair burden sharing. Based on these principles and responsibilities, the law stipulates the policy instruments of Japanese environmental policy as follows:

- environmental consideration in policy formulation;
- formulation of the Basic Environmental Plan which sets out the outline of long-term environmental policy (this plan was formulated in December 1994);
- environmental impact assessment for development projects;
- economic measures to encourage activities for reduction of environmental burdens;
- improvement of social infrastructure such as sewerage systems, transportation systems, etc.;
- protection of environmental activities by corporations, citizens, and non-governmental organizations (NGOs), as well as environmental education and dissemination of information;

- advancement of science and technology related to environmental protection;
- international cooperation for global environmental conservation.

The Basic Environmental Plan

In accordance with the provisions of the Basic Environment Law, the Basic Environmental Plan was implemented by the Japanese Cabinet in December 1994. The plan provides not only the basis for government policies but also details the roles and activities expected to be carried out by local governments, corporations, NGOs, and the public for achieving the plan's objectives. The plan sets long-term objectives for environmental policies through to the middle of the twenty-first century, which are:

- building a socio-economic system which fosters a sound materials cycle;
- harmonious coexistence between humankind and nature;
- participation by all sectors of society;
- promotion of international activities.

Air Pollution Control Law

The Air Pollution Control Law introduced new legal measures in June 1968. The so-called "Pollution Diet" in 1970 added numerous amendments to this law to meet social and administrative requirements. To date, this revised law has played a major role in preventing air pollution in Japan. The Air Pollution Control Law provides a broad basis for controlling emissions of air pollutants from both stationary and mobile sources, as shown in Table 8.3. Regarding the control of stationary sources, it enables the government to set up emission standards and implement them through registration and inspection. The emission standards are established by the national government although they can be replaced, except for sulphur oxides, by more stringent standards set by prefectural by-laws. As presented in Table 8.3, most of the implementation of this law is delegated to local governments.

In addition to the emission standards, a 1974 amendment of the law introduced a new regulation system called "total emission control" for sulphur oxides. In 1981, the same system was applied to nitrogen oxides. Under the system, prefectural governors determine the maximum amount of emission that would be tolerated within each area regulated, draw up programmes to reduce total emissions to the level determined, and set standards for total volume of emissions on a plant-by-plant basis. This system was introduced in areas where emission standards were not sufficient to maintain the environment quality standards set by the national government.

Table 8.3 Framework for Air Pollution Control Law

Overall purpose
- To protect public health and preserve the human living environment with respect to air pollution.
- To help victims of air-pollution-related damage by providing a liability regime for health damage caused by air pollution from business activities.

Legislative area	Elements	Implications
Soot and smoke (Chapter 2)	Sulphur oxides Soot and dust Toxic substances	• obligation of notification and restrictions on implementation of plans for the emitting facility • modification of proposed plans • obligation to measure the volume of soot and smoke • regulations on use of fuel
	Designated soot and dust	• for certain factories, development of a total mass reduction plan • total/specific mass emission control standards • fuel standards
	Specific substances	• measurements in case of accidents • obligation of notification of the situation of accidents
	Emergency measures	• ask persons responsible for emitting soot and smoke to cooperate in reducing the level of emissions • order such persons to take necessary measures
Particulates (Chapter 2-2)	General particulates	• obligation of notification • standards relating to structure, operation, and management • obligation to observe the standards
	Designated particulates	• obligation of notification • standards on the borderline • obligation to observe the standards
Hazardous air pollutants (Chapter 2-3)	*Guiding principles* • corporations – undertake measures to reduce their own emissions • the state – evaluate the state of air pollution, enrich scientific knowledge, evaluate and announce officially the health risk of hazardous air pollutants, gather and spread appropriate remediation technologies • local governments – evaluate the state of air pollution in their territories, provide corporations and citizens with information	

Table 8.3 (cont.)

Legislative area	Elements	Implications
Motor vehicle exhausts (Chapter 3)	Exhaust gas from motor vehicles	• maximum permissible limits: responsibility of the Ministry of Transport (MOT) • maximum permissible limits on quality of automobile fuel: responsibility of MOT and MITI
	Emergency measures	• ask for cooperation in reducing the level of emissions • demand the Prefectural Public Safety Commission to undertake measures
Monitoring the level of air pollution (Chapter 4)	• monitoring and surveillance from time to time by the prefectural governor • public announcement by the governor of the prefecture	
Compensation for damages (Chapter 4-2)		
Others	• report and inspection • demand for data • assistance by the state • promotion of research	

Source: Environment Agency of Japan (1998)

Automobile exhausts are also controlled by the Air Pollution Control Law. However, exhaust standards are not good enough to meet the environment quality standard for NO_2 in urban areas because of the rapid increase of transportation. In 1992, a new law was passed to address NO_x pollution problems in urban areas. This law, called the Law Concerning Special Measures for the Total Emission Reduction of NO_x From Automobiles, provides special measures, including restrictions on the use of designated types of vehicles and guidance to automobile users for reducing NO_x emissions.

In 1989 the Air Pollution Control Law was amended to control asbestos dust from asbestos manufacturing factories. In 1995 the law was amended to control the quality of automobile fuels, and in 1996 the law was again amended to control various hazardous air pollutants such as benzene, to control automobile exhaust from small two-wheeled motor vehicles, to control asbestos at the time of demolition of buildings, and to strengthen measures when there is an accident in factories, etc.

Environmental quality standards for air pollution

Environmental quality standards for air pollution are targets for the protection of air quality, as shown in Table 8.4. These standards are set under the Basic Environment Law and based on scientific knowledge which shows the effect of pollutants on human health under various exposure conditions. As a general principle, policies concerning energy, industry, and transportation should be developed while integrating environmental aspects. Prior to enactment of the Basic Law for Environmental Pollution Control in 1967, which was amended to the current Basic Environment Law in 1994, emission controls had been applied on the basis of individual sources of pollution; this was later found to be largely ineffective because of the rapid increase of pollution sources. To enhance a comprehensive approach to solving the increasing air pollution problems, the Basic Law for Environmental Pollution Control empowered the government to set up the environmental quality standards, including air quality, for the protection of human health and the conservation of the living environment.

The former Environment Agency of Japan set environmental quality standards for five traditional air pollutants: sulphur dioxide, carbon monoxide, nitrogen dioxide, suspended particulate matter, and photochemical oxidants. There is also growing concern about "hazardous air pollutants" (HAPs), such as benzene, trichloroethylene, and tetrachloroethylene. Their concentrations are much smaller than those of traditional air pollutants, but they can cause cancer and other health effects through long-term exposure. In 1997, based on reports submitted by the Central Environment Council, the Environment Agency set environmental quality standards for benzene, trichloroethylene, and tetrachloroethylene. In addition to environmental quality standards, there are two guidelines concerning ambient air quality in Japan. One was set for non-methane hydrocarbons in 1976 to attain environmental quality standards for photochemical oxidants, and the other was set for dioxins in 1997 as a goal to decrease their concentration.

Regulatory mechanisms for stationary and mobile sources

Regulatory mechanisms for stationary sources

Categories of facilities under control

The Air Pollution Control Law specifies three types of facilities to be controlled: "soot and smoke emitting facility", "general dust discharging

Table 8.4 Environmental quality standards in Japan (air quality)

Substance	Environmental conditions	Measuring method
Sulphur dioxide	Daily average of hourly values shall not exceed 0.04 ppm, and hourly values shall not exceed 0.1 ppm (notification on 16 May 1973)	Conductometric method or ultraviolet fluorescence method
Carbon monoxide	Daily average of hourly values shall not exceed 10 ppm, and average of hourly values in eight consecutive hours shall not exceed 20 ppm (notification on 8 May 1973)	Non-dispersive infrared analyser method
Suspended particulate matter[a]	Daily average of hourly values shall not exceed 0.10 mg/m^3, and hourly values shall not exceed 0.20 mg/m^3 (notification on 8 May 1973)	Weight concentration measuring methods based on filtration collection, or light-scattering method, or piezo-electric microbalance method, or beta-ray attenuation method combining the values of the above methods
Nitrogen dioxide	Daily average of hourly values shall be within the range from 0.04 ppm to 0.06 ppm or below (notification on 11 July 1973)	Colorimetry employing Saltzman reagent (with Saltzman's coefficient being 0.84) or chemiluminescent method using ozone
Photochemical oxidants[b]	Hourly values shall not exceed 0.06 ppm (notification on 8 May 1973)	Absorption spectrophotometry using neutral potassium iodide solution or coulometry, or ultraviolet absorption spectrometry, or chemiluminescent methods using ethylene
Benzene Trichloroethylene Tetrachloroethylene	Annual average shall not exceed 0.003 mg/m^3 Annual average shall not exceed 0.2 mg/m^3	Preference method: gas chromatograph-mass spectrometer (sample gas should be collected with canister or tube) or its equivalent method

a. Suspended particulate matter is defined as airborne particles with diameter smaller than or equal to 10 μm.

b. Photochemical oxidants are oxidizing substances such as ozone and peroxiacetyl nitrate produced by photochemical reactions (only those capable of isolating iodine from neutral potassium iodide, excluding nitrogen dioxide).

facility", and "specific dust discharging facility" (Environment Agency of Japan 1999); these are presented in detail in Table 8.5. A "soot and smoke emitting facility" means any designated facility installed in an industrial plant or business establishment which generates or emits air pollutants such as SO_x and NO_x. The types of facility have to date been increased to 32, such as boilers, gas generating furnaces, blast furnaces, etc. Recently designated facilities have been added, such as gas turbine and diesel engines in 1988, and gas engines and gasoline engines in 1991. The number of facilities under control reached 203,000 nationwide in 1996. A "general dust discharging facility" means any designated facility which discharges or scatters dust as a result of mechanical treatments, such as crushing and selection of materials. Five types of facilities are specified, including conveyors, crushers, and sieves, and their number amounted to 59,000 nationwide in 1996. A "specific dust discharging facility" means any designated facility which discharges or scatters specific dust; asbestos was designated as a specific dust in 1989. Nine types of facilities, including grinders and cutters, are designated and their number amounted to 2,141 nationwide in 1996.

Reporting and inspection of facilities

The Air Pollution Control Law orders any person who plans to establish or modify any of the abovementioned facilities to report the necessary information to prefectural governors in order to examine whether the plans will meet control standards. Prefectural governors are empowered by the law to inspect the operating facility for the implementation of emission standards, etc.

Air pollutants under regulatory control

Sulphur oxides

Because desulphurization technology had not been fully developed when the emission standard for sulphur oxides was set up, the emission standard was determined as permissible emission levels in proportion to the square of the height of smokestacks, in line with the idea that slight reductions in air pollution can be achieved with taller smokestacks. Although this regulation method did contribute to ease the pollution from individual sources, it was not good enough to improve the pollution in heavily industrialized areas. It also contributed to spreading the pollution over a larger area. Emission standards for sulphur oxides vary with the location and the stack height of emission sources, but are uniform regardless of the type of fuel used.

There is also a fuel standard regarding sulphur content under the Air Pollution Control Law. The original fuel standard, which was set in 1971,

Table 8.5 Regulatory measures against air pollutants emitted from factories and business sites and the outline of regulation

Name of substance	Main form of generation	Form and outline of regulation
Soot and smoke		
Sulphur oxides (SO_x)	Combustion of fuel and minerals in boilers and waste incinerators	1. The regulatory value(quantity) is set according to the height of the exhaust outlet and the value of the consent, K, designated at each area. Permissible emission (Nm^3/h) = K × 10–4 × He2 General emission standard: K = 3.0 ~ 17.5 Specific emission standard: K = 1.17 ~ 2.34 2. Fuel use standard. Sulphur in fuel is set at each area Sulphur content: under 0.5 ~ 1.2% 3. Regulation of total emissions: set at each area/factory based on the total emission reduction plan.
Soot and dust		Emission standards for each facility. General emission standard: 0.04 ~ 0.7 g/Nm^3 Specific emission standard: 0.03 ~ 0.2 g/Nm^3
Harmful substances		
Cadmium (Cd) Cadmium compounds	Combustion or chemical treatment at copper, zinc, or lead refinery	Emission standards for each facility. 1.0 mg/Nm^3
Chlorine (Cl_2) Hydrogen Chloride (HCl)	Combustion or chemical treatment at chemical product reaction facilities or a waste incinerator	Emission standards for each facility. Chlorine: 30 mg/Nm^3 Hydrogen chloride: 80,700 mg/Nm^3

Fluorine (F) Hydrogen fluorine (HF)	Processes at electric furnaces for aluminum refinement or fusion furnaces for glass production	Emission standards for each facility. $1.0 \sim 20$ mg/Nm^3
Lead (Pb) Lead compounds	Combustion or chemical treatment at copper, zinc, or lead refinery	Emission standards for each facility. $10 \sim 30$ mg/Nm^3
Nitrogen oxides (NO_x)	Combustion, synthesis, or degradation in a boiler or a waste incinerator	1. Emission standards for each facility/scale. New facilities: $60 \sim 400$ ppm Existing facilities: $130 \sim 600$ ppm 2. Regulation of total emission: set at each area/factory based on the total emission reduction plan.
Dust		
General dust	Comminution/selection, mechanical treatment, accumulation at a sieve or an accumulation site	• Standard regarding the structure, use and administration of facilities. • Installation of a dust collector; a dust protection cover or hood; water spray.
Designated dust (asbestos)	Comminution of asbestos with a cutter, mixing and other mechanical treatment	Site boundary standard at the business site. Concentration 10/litre
	Dismantling or repair of a building with sprayed asbestos	Standard regarding removal/enclosure/containment at the time of dismantling the building.
Designated substances (28 harmful substances)	Accident such as disorder/damage at a factory	Specification of measurements at the time of an accident. Duty of the company for restoration; report to the prefectural governor.

Table 8.5 (cont.)

Name of substance	Main form of generation	Form and outline of regulation
Harmful air pollutants (substances suspected to have chronic health effects)	234 substances (groups): of these, 22 are priority substances	Specification of duties for each subject, such as the accumulation of knowledge. Voluntary tackling of companies and people such as the reduction of emissions, the development of scientific knowledge by the nation, grasping the condition of pollution by self-governing bodies.
Benzene	Facilities for drying benzene	Suppression standard for each facility/scale. New facilities: 50 ~ 600 mg/Nm^3 Existing facilities: 100 ~ 1,500 mg/Nm^3
Trichloroethylene	Facilities for cleaning with trichloroethylene	Suppression standard for each facility/scale. New facilities: 150 ~ 300 mg/Nm^3 Existing facilities: 300 ~ 500 mg/Nm^3
Tetrachloroethylene	Dry-cleaning with tetrachloroethylene	Suppression standard for each facility/scale. New facilities: 150 ~ 300 mg/Nm^3 Existing facilities: 300 ~ 500 mg/Nm^3
Dioxins	Electric furnaces for purifying steel, incinerators for industrial wastes	Suppression standard for each facility/scale. New facilities: 0.1 ~ 5 ng/Nm^3 Existing facilities: 1.0 ~ 10 ng/Nm^3

Source: Environment Agency of Japan (1999)

was upgraded in 1976 to a sulphur content ranging from 0.5 to 1.2 per cent. This standard is applied to factories and business offices located in 14 areas which are specified by Cabinet Order: Sapporo, Asahikawa, Sendai, Chiba, central Tokyo, Yokohama, Kawasaki, Nagoya, Kyoto, Osaka, Kobe, Amagasaki, Hiroshima, and Fukuoka.

By a June 1974 amendment to the Air Pollution Control Law, a system of control on the total volume of emissions within a given area was introduced in an attempt to enable prompt improvement of air quality. The applicable regions are those where plants and businesses are clustered and environment quality standards will not be easily attained by application of existing regulations alone. For such a region, a maximum total amount of pollutant emission is computed with a scientific pollution forecasting method which takes into account the meteorological and topographic features of the region as well as the conditions of the emission sources located therein (Bradstreet 1995).

Nitrogen oxides

NO_x are a set of pollutants that demand the closest attention because not only are they harmful to human health but they are also responsible for causing photochemical air pollution (smog). A nitrogen oxides emission standard was set up as permissible concentration limits varying with the types and scales of facilities. Though it is designed as a uniform standard nationwide, local governments are allowed to make it more stringent if this is necessary to meet the environment quality standard. The emission standard for nitrogen oxides was first introduced in August 1973 (first regulation). After that, the standards were strengthened and the types of facility under control were added in December 1975 (second regulation), in June 1977 (third regulation), in August 1979 (fourth regulation), and in September 1983 (fifth regulation). The fifth regulation was made in response to changes in energy supply from oil to solid fuels such as coal, which generate more nitrogen oxides, and progress in combustion technology to reduce NO_x emissions.

Though sources of NO_x are both stationary and mobile, only stationary sources are subjected to total emission control under the Air Pollution Control Law. In June 1981 this control system was introduced to the three areas of Tokyo, Kanagawa, and Osaka. When considering application of a total emission standard to a designated establishment, a stricter standard was generally set for larger establishments in view of their superior ability to take measures. It is also possible in certain cases, as when a designated establishment is being enlarged or a new designated establishment is being constructed, to apply a special total emission standard which is more stringent than the standard applied to existing establishments.

Suspended particulate matter

Particulate matter within air is classified largely into "falling dust" and "suspended dust". The suspended dust is further classified into "suspended particulate matter", in which the grain diameter is 10 micrometres or less according to the environmental standard, and others. The Air Pollution Control Law classifies dust according to the process by which the dust was generated. Particulate emission standards represent the permissible limits of particulates contained in the emission gas released into the air from the outlets of emitting facilities. They vary with types and scales of facilities. The ordinary emission standards are the ones which apply uniformly throughout the country, while the special emission standards are those applied to facilities which are newly installed in specified areas where air pollution is a greater problem. The latter are more stringent than the former.

Concerning suspended particulate matter, the environmental standard was established in January 1972, but effective measures have not been determined yet because of the complicated generation mechanism. Suspended particulate matter is generated as secondary particles that grow from gaseous matter such as sulphur oxides as a result of physical and chemical changes in the atmosphere. Various investigations and field surveys on the sources of suspended particulate matter have been conducted to develop methods of estimating pollution levels and reducing them effectively.

Dust

Emission standards for soot and dust were established by type of facility and its scale in 1971. Standards cover the structure, use, and management of general dust discharging facilities, such as places for dumping minerals, conveyors, etc. The measures include promotion of appropriate combustion methods and installation of dust collecters.

Asbestos was designated as specific dust by a revision to the Air Pollution Control Law in 1989. Control standards for asbestos (defined as less than 10 fibres per litre at the boundary of a factory) are applied to any factory with facilities to manufacture asbestos products. Also asbestos discharging standards were developed in 1997 for buildings with more than 50 m^2 of asbestos sprayed area. These specify management standards for activities in these buildings, such as use of water sprinklers. The Environment Agency of Japan (EAJ) conducts investigations on development of alternative products to replace asbestos.

Dust cased by automobiles with studded tyres not only worsens the living environment but also has a bad influence on human health. In June 1990 the Studded Tyres Regulation Law was enforced. The law describes the responsibility of people to strive to prevent the dust caused by studded tyres and the restrictions on the use of these tyres in areas specified by the director-general of the EAJ. As of April 1996, 817 municipalities in 18 prefectures have been designated as restricted areas. The law also says that national and local governments must strive to promote and enforce measures to prevent the generation of dust caused by studded tyres.

Hazardous substances

In addition to NO_X, the Air Pollution Control Law designated four groups of substances as "hazardous substances" generated from soot and smoke emitting facilities, and stipulates control of their emission levels: cadmium and its compounds; chlorine and hydrogen chloride; fluorine, hydrogen fluoride, and silicon fluoride; and lead and its compounds. Standards are set for each of the four groups of hazardous substances only, and are applicable to a very limited number of soot and smoke emitting facilities because the generation of hazardous substances is related to certain materials. In response to recent dioxin air pollution issues concerning waste incinerators, the Reinforcing Reform of the Soot and Dust Regulations regarding waste incinerators was enforced in July 1998. Based on the Five-Year Plan on Dioxins Measures made in August 1997, measures including generation source controls and comprehensive monitoring surveys will be taken to promote the suppression of dioxin emissions (Environment Agency of Japan 2000).

Photochemical oxidants

The concentrations of photochemical oxidants are still higher than the environmental quality standard in most areas. For emergency measures, the EAJ observes meteorological conditions at four points within Tokyo Bay and Osaka Bay, where levels of photochemical oxidants frequently become high during summer. Based on this meteorological information, the issue of warnings is decided. The agency supples the data to the relevant local governments. At 19 meteorological observation stations, the Meteorological Agency analyses and forecasts weather conditions and announces meteorological information related to photochemical air pollution to the public. Based on this information from monitoring, local governments issue warnings or alarms according to the procedure on emergency measures against photochemical oxidants; these are outlined in Table 8.6. At the same time, local governments request persons responsible for the emission of soot and smoke to reduce the emission of air pollutants, and drivers to refrain from non-essential use of vehicles. They also carry out public information activities and provide public health measures.

Table 8.6 Stipulations for an emergency situation

	The Law: Article 23, Paragraph 1 (warning or alarm)		The Law: Article 23, Paragraph 4 (serious emergency)	
Sulphur oxides	0.2 ppm	3 hours	0.5 ppm	3 hours
	0.2 ppm	2 hours	0.7 ppm	2 hours
	0.5 ppm	1 hour		
	0.15 ppm	48 hours average		
Suspended particulates	2.0 mg/m^3	2 hours	3.0 mg/m^3	3 hours
Carbon monoxide	30 ppm	1 hour	50 ppm	1 hour
Nitrogen dioxide	0.5 ppm	1 hour	1 ppm	1 hour
Oxidants	0.12 ppm	1 hour	0.4 ppm	1 hour

Source: Environment Agency of Japan (1999)

Regulatory measures for automobile exhausts (mobile sources)

Regulatory system under the Air Pollution Control Law

The Air Pollution Control Law empowered the former Environment Agency of Japan to establish the permissible limits of motor vehicle exhaust, or the exhaust standards. Permissible limits have been set up for four pollutants; CO, HC, NO_x, and diesel smoke and particulate matter. In order to ensure the permissible limits, the law requires the Minister of Transport to take necessary measures under the Road Vehicle Act, which obliges any motor vehicle to satisfy the permissible limits of exhaust in its initial or continuation inspection or type-approval test.

The automobile exhaust was first regulated in 1966 for CO. Since then, regulation has been strengthened on a step-by-step basis. It now covers CO, HC, NO_x and particulate (diesel black smoke). The existing exhaust emissions standard for NO_x was applied to gasoline and LPG-powered automobiles in 1978. The regulation for diesel-powered automobiles has been strengthened several times. As the result, the regulation has become one of the most stringent in the world. Considering the serious pollution of NO_x, the Central Council for Environmental Pollution Control issued recommendations to reduce further the exhaust from automobiles in December 1989. The recommendations included:

- reducing NO_x emissions by 30–60 per cent from diesel-powered automobiles;
- reducing the emission of particulate by more than 60 per cent from diesel-powered automobiles;

Table 8.7 Permissible limits for automobile fuels (as of October 1995)

Type of fuel	Character of fuel or materials contained in the fuel	Permissible limits
Gasoline	Lead	Not detected
	Sulphur	No more than 1% in weight
	Benzene	No more than 5% in volume (1% after 1998)
	Methyl tertiary butyl ether (MTBE)	No more than 7% in volume
Diesel fuel	Sulphur	No more than 2% in weight
	Setane index	No more than 45
	T90	No more than 360 °C

Note: T90 stands for distillation temperature at which 90 per cent of fuel has been evaporated.
Source: Environment Agency of Japan (1999)

- reducing the sulphur content in diesel fuel;
- reviewing the testing mode for exhausts.

The Council set up a short-term target (within five years) and long-term time scale (within 10 years) to implement the recommendations. The short-term target was realized in 1991 and came into force from 1992 to 1994. Since it was regarded as possible to realize the long-term target by 1999, the director-general of the EAJ asked the Central Council for Environmental Pollution Control in May 1996 to consider further strengthening of exhaust standards.

In 1995 the Air Pollution Control Law was amended to control the quality of automobile fuels, because poor-quality automobile fuels may worsen air pollution in Japan. Based on this amendment, the former Environment Agency of Japan could decide permissible limits on the amount of substances in automobile fuels (Table 8.7). In order to ensure the permissible limits, the law requires the Minister of International Trade and Industry to take necessary measures under the Law on the Quality Control of Gasoline and Other Fuels.

On 18 October 1996 the Central Environment Council submitted an interim recommendation which dealt with measures for hazardous air pollutants:

- to reduce the HC emission for two-wheeled vehicles from 1998 to 1999 (presented in Table 8.8);
- to reduce the HC emission for gasoline/LPG motor vehicles in 1998 (presented in Table 8.8);
- to reduce the benzene content in gasoline in 1999 (presented in Table 8.7).

Table 8.8 Target values for permissible limits (1996)

Category of motor vehicles	Target values of permissible limits (mean values)			Measurement method
	Carbon monoxide	Hydrocarbons	Nitrogen oxides	
Small-sized motor vehicles (two-wheeled motor vehicles)				
With 4-stroke engine	13.0 g/km	2.00 g/km	0.30 g/km	Two-wheeled vehicle measurement mode
With 2-stroke engine	8.00 g/km	3.00 g/km	0.10 /km	Two-wheeled vehicle measurement mode
Mini-sized motor vehicles fueled by gasoline or LPG[1]	6.50 g/km	0.25 g/km	0.25 g/km	10.15-mode[3]
Ordinary-sized and small-sized motor vehicles fueled by gasoline or LPG[2]				
Gross vehicle weight of 1,700 kg, but no more than 2,500 kg or less	6.50 g/km	0.25 g/km	0.40 g/km	10.15-mode
Gross vehicle weight in excess of 2,500 kg	51.0 g/kwh	1.80 g/kwh	4.50 g/kwh	10.15-mode

Notes:
1. Excluding those used exclusively for carriage of passengers and those with 2-stroke engines.
2. Excluding those used exclusively for carriage of passengers with a capacity of 10 persons or less.
3. The 10.15-mode represents a typical driving pattern in urban areas.
Source: Environment Agency of Japan (1999)

Table 8.9 Target values for permissible limits for diesel off-road vehicles (1997)

Category of motor vehicles	Nitrogen oxides	Hydrocarbons	Carbon monoxide	Particulate matter
Those with a rated output of 19 kW or more, but less than 37 kW	8.0 g/kWh	1.5 g/kWh	5.0 g/kWh	8.0 g/kWh
Those with a rated output of 37 kW or more, but less than 75 kW	7.0 g/kWh	1.3 g/kWh	5.0 g/kWh	0.4 g/kWh
Those with a rated output of 75 kW or more, but less than 130 kW	6.0 g/kWh	1.0 g/kWh	5.0 g/kWh	0.3 g/kWh
Those with a rated output of 130 kW or more, but less than 560 kW	6.0 g/kWh	1.0 g/kWh	3.5 g/kWh	0.2 g/kWh

Source: Environment Agency of Japan (1999)

On 21 November 1997 the Council submitted a second set of recommendations:

- to reduce the NO_x, HC, and CO emissions by 70 per cent from exhaust pipes of gasoline/LPG motor vehicles from 2000 to 2002;
- to extend the durability running distance and to install the OBD (on-board diagnosis) system for gasoline/LPG motor vehicles from 2000 to 2002;
- to reduce fuel evaporative emissions from gasoline motor vehicles from 2000 to 2002;
- to reduce the NO_x and particulate matter emissions from diesel off-road vehicles in 2004 (presented in Table 8.9).

Regulation for automobile NO_x emissions

Continued efforts to reduce the emissions from each vehicle have been offset, particularly in major cities, by the increase in transportation volume. It is especially obvious in the case of NO_x. The increase in the number of diesel vehicles which discharge large amounts of NO_x is the major reason for NO_x pollution problems. In order to cope with the NO_x pollution caused by automobiles, a new law was promulgated in June 1992, namely the Law Concerning Special Measures for Total Emission Reduction of Nitrogen Oxides from Automobiles in Specified Areas, or the Automobile NO_x Law. This law provides comprehensive measures on the vehicle NO_x problem, including restrictions on the use of designated types of vehicles in specified areas and guidance for vehicle users to reduce NO_x emissions. The specified areas, designated in November 1992, are parts of Tokyo, Kanagawa, Saitama, Chiba, Osaka, and Hyogo, where buses and trucks must meet the specified vehicle emissions standards. In addition, each prefectural governor in the specified areas formulated the NO_x Emission Reduction Plan in November 1993. This plan shows target levels of NO_x emissions in each prefecture to achieve the environment quality standard for NO_2, as well as various measures such as promoting low-emission vehicles, traffic demand management, a construction plan for distribution centres for commercial goods, etc.

Economic measures

National economic incentives have been provided through the Japan Environment Corporation (JEC), which the Ministry of Health and Welfare founded in 1965. The JEC provides low-interest financing to businesses for installing pollution control measures in existing facilities, and to industrial companies as assistance in relocating away from residential areas. For small and medium-size businesses, the JEC provides special loans for up to 80 per cent of pollution control costs for such businesses, while covering 30 per cent for larger businesses. Two other institutions, the Small Businesses Promotion Council (SBPC) and the Small and Medium Enterprise Agency, are also important sources of low-cost financing to help industries to comply with environmental standards. While the policy emphasizes assisting environmentally sound investment, the polluter pays principle (PPP) was recognized in the Basic Law on the Environment in 1993. Though the principle has been adapted, one noteworthy application of the PPP is the Law for the Compensation of Pollution-Related Health Injuries in 1973. The law introduces taxes on emissions and effluent, the proceeds of which go into a fund for compensation to pollution victims.

Monitoring and information dissemination system

The constant monitoring of air pollution is vital to understanding the achievement of environmental standards, to establishing control measures against air pollution, and to evaluating the improvement of the monitoring system. Since the establishment of the EAJ, environmental monitoring and information have been gradually strengthened along with the development of environmental policy. These data have been effectively utilized for policy decision-making and plan formation in environmental pollution control programmes, pollution control administration, and environmental information assessment (Hashimoto 1985).

National atmospheric monitoring stations

Three sorts of monitoring stations are placed to evaluate the state of air pollution on a nationwide scale and to obtain basic data necessary to promote measures for atmospheric preservation:

- national air pollution monitoring stations (15 stations), to understand the status of air pollution in major areas;
- national environmental atmospheric monitoring stations (eight stations), to understand levels of pollution in areas that have not yet been polluted in major plains;
- national roadside air pollution monitoring stations (four stations), to monitor the levels of roadside air pollution – three stations are placed within the Tokyo metropolis, and one in Maebashi city, Gunma prefecture.

Regarding national air pollution monitoring stations and national environmental atmospheric monitoring stations, each monitoring station has various types of measuring equipment to monitor constantly the level of pollutants such as sulphur dioxide, nitrogen oxides, suspended particulate matter, etc. Analysis of metal composition of suspended particulate matter and suspended fragmentary dust is also conducted. Some of the national environmental atmospheric monitoring stations perform measurements of hydrogen sulphide, ozone, mercury, etc.

National acid deposition monitoring stations

Acid deposition measurement equipment was installed in 23 atmospheric monitoring stations in 1987. Since 1989, acid deposition monitoring stations have been built throughout the country to perform analysis of ph levels and the composition of precipitation. In 1994, three additional stations started their services.

Local air pollution monitoring system

In local areas, the monitoring of air pollution levels is conducted under the supervision of the governor of each prefecture or metropolis, or the mayor of each administrative ordinance designated city, in conformity with the Air Pollution Control Law. The installation of telemetric devices that constantly monitor levels of sulphur dioxide and the amount of fuel used, etc. has been in progress. These data are transmitted to the central monitoring centre. Governmental support to improve the measurement equipment for the monitoring conducted by prefectures and metropolises has been provided. The government also encourages improvement of monitoring systems to achieve efficiency, and to raise the standard of measuring technology (Broadbent 1998). In order to comprehend the actual state of hazardous chemical substances, improvement of the facilities of the Local Institute for Environmental Pollution Control has been promoted by the government as well.

Information dissemination system

Air quality monitoring data are provided via homepages of the Environmental Information Centre (EIC) and the National Institute for Environmental Studies (NIES). The EIC provides various functions and services related to environmental information via databases, a library, and a computer system.

Environmental databases

The environmental databases contain three components. The first is a facility for monitoring data files. A wide range of numerical environmental data are necessary for both environmental research and policy development, implementation, and enforcement. The EIC has compiled, processed, stored, and provided access (in computer-accessible form) to data files of air quality and water quality monitoring data which are transmitted by local governments to the Environment Agency under the Air Pollution Control Law and the Water Pollution Control Law. These data files are provided to outside users, including other governmental organizations and laboratories. A duplication service for use by the general public is also available for some files; in addition data files are exchanged with other governmental organizations. Air quality monitoring data files contain the following parameters (figures in parentheses show the year data collection began):

- the hourly ambient air quality data file (1976)
- the hourly ambient air quality data file (national station) (1976)
- the monthly and yearly ambient air quality data file (1997)
- the ambient air monitoring stations attribute data file (1996).

The second component concerns the natural environment, providing basic reference materials which facilitate both understanding of present conditions and forecasting of changes in the natural environment. Development of a general reference system for the natural environment began in 1991. Since 1995 a system to provide database access by personal computers (P-GREEN) has been developed based on previously recorded results and data. P-GREEN is available on Windows-based PCs, enabling graphical display and user-friendly operation.

The final component provides environmental information sources. Surveys of environmental information have been in progress since 1992, resulting in a directory of information sources in a form widely accessible to the general public. The surveys – including information about where and in what mode environmental information is being accumulated (environmental information sources), and explanations of laws, treaties, and terms concerning the environment – were compiled on floppy disks and are being distributed to the general public through a public corporation and through NIES and EIC web servers.

EIC net

Under the Basic Environment Law, the government provides information on environmental conservation in order to promote education and learning and to contribute to the activities voluntarily conducted by private bodies. The EIC provides important information which will be useful for environmental conservation and environmental awareness among Japanese citizens through the Environmental Information Providing System, via a personal computer and facsimile telecommunication system. The EIC also provides a directory database of information owned by governmental organizations, prefectural governments, private bodies, etc. on this system.

In March 1996 the EIC established a computer communication system for the general public, called the Environmental Information and Communication Network (EIC net), to support information exchange on environmental conservation activities such as events and techniques among citizens and private bodies, etc. This system is available only in Japanese, via telephone, internet, or the Value-Added Network (VAN). In January 1997 an EIC net web server was also established. In December 1997, the EIC started to provide environmental information by facsimile, covering information such as:

- directory database of environmental information;
- research and statistical data concerning environmental issues;
- environmental administrative information (white papers, laws, plans, etc.);
- directory database of environmental publications;
- information on environmental conservation activities, etc.

Role of NGOs

When air pollution first became a problem, Japanese NGOs did not possess sufficient knowledge of the problem and were not able to propose any response measures. At the same time, the response from industry and government was also insufficient. This led to a relatively insignificant role for the NGOs during the 1960s and 1970s (Sotome 1997; Schreurs 1996).

This situation has changed in recent years; according to a recent survey by the Economic Planning Agency, the total number of Japanese NGOs is now about 8,500. Among them, 1,558 NGOs (36.9 per cent) focus on the issues of air pollution and air quality improvement. In particular, the Japan Pollution Victims Association (Zenkoku kogai kanjya-no kai rengo kai) and Citizens' Alliance for Saving the Atomosphere and the Earth – CASA (Chikyukankyo to Taikiosen wo kangaeru zenkoku shimin kaigi) play active roles in this field (Japan Environmental Corporation 1998). The Japan Pollution Victims Association is a network of victims of air pollution, and CASA conducts research on air pollution, organizes symposia, collects and provides information, and creates networks with NGOs elsewhere in the world on air pollution and global environmental issues.

CASA was established in 1988, for the purpose of protecting both the local and global environments through solidarity with both overseas and local Japanese NGOs. During the Nishiyodogawa pollution trial in Osaka, CASA helped settle the case out of court and made significant contributions to setting a vehicle exhaust pollution limit. It was a historical precedent in Japan. In June every year, CASA participates in the National Action of Pollution Victims campaign. CASA has been advocating the establishment of a more effective Basic Environmental Law and an environmental assessment system through public hearings. Along with other associations in the Kinki region, CASA also holds symposia and research meetings with officers from environmental agencies. CASA strongly encourages public participation in the decision-making process on environmental issues, particularly in the early stages of discussion.

However, NGOs in Japan are relativity small in terms of scale, finance, and manpower compared with those in Europe and North America. In addition, difficulties in gaining tax-exempt status, insufficient access to environmental information, and poor participation in the decision-making process by serving on advisory committees (*shingikai*) to the ministries can be mentioned (Imura 1993). Public access to environmental information has been limited, though the Basic Law for the Environment stipulates that the government should provide the necessary environmental information. However, there are no specific guidelines for distributing the information to the public; this lack of access results from the limited power of NGOs to influence policy. Their funding is very limited because their membership is relatively small and few organizations qualify for tax-deductible donations. In addition, NGOs do not have the legal standing to bring environmental cases to court.

But compared with past, Japanese NGOs can utilize abundant information. It will be important in the future for NGOs' activities to contribute to swift resolution of problems based upon either accurate information in the public domain or their own additional investigations and research (Committee on Japan's Experience in the Battle against Air Pollution 1997).

The Japanese government has been putting considerable effort into providing support for NGOs, mainly through the Ministry of Foreign Affairs (MOFA). Since 1989, MOFA has been operating a subsidy system for NGO projects which provides a 50 per cent subsidy of between ¥500,000 and ¥1.5 million per project; the total budget has increased from ¥100 million in 1989 to ¥1.2 billion in 1997. Other ministries and agencies, such as the Ministry of Post and Telecommunications, the Ministry of Agriculture, Forestry, and Fisheries, and the EAJ also have NGO assistance budgets, thought the figures are relatively small. To support NGOs, the Global Environmental Information Centre (GEIC) and the Japan Fund for Global Environment were established.

Global Environmental Information Centre

As a joint project between the EAJ and the United Nations University, the GEIC was opened in October 1996. The GEIC is aimed at assisting governments, international organizations, NGOs, consumers, business operators, and all other constituents of society in promoting environmental activities based on their partnership. It is carrying out dissemination of environmental information, seminars, exhibitions, and international symposia, including the promotion of cooperative efforts among the United Nations, governments, and NGOs, etc.

The Japan Fund for Global Environment

The Japan Fund for Global Environment was established within the Japan Environmental Corporation in May 1993 with a view to extending financial, informational, educational, and training assistance to NGOs inside and outside Japan. The fund is endowed not only by the national government but also by citizens and corporations. Two main projects have been conducted through this fund: one providing assistance for environmental conservation activities of private organizations; the other supporting dissemination of information necessary for promoting activities of private organizations as well as education and training. In 1997 a second global partnership programme was inaugurated in order to help organizing a worldwide network of NGOs and to foster partnership among countries in the Asia Pacific region for environmental conservation activities. Under this programme, priority is assigned to these activities in the allocation of its funds.

REFERENCES

Abe, Y. and T. Awaji (eds). 1995. *Kankyo Ho (Environmental Laws in Japan)*. Tokyo: Yuhikaku Books.

Bradstreet, W. J. 1995. *Hazardous Air Pollutants: Assessment, Liabilities, and Regulatory Compliance*. Park Ridge: Noyes Publications.

Broadbent, J. 1998. *Environmental Politics in Japan: Networks of Power and Protest*. Cambridge: Cambridge University Press.

Committee on Japan's Experience in the Battle against Air Pollution (ed.). 1997. *Japan's Experience in the Battle against Air Pollution: Working Toward Sustainable Development*. Tokyo: Pollution-Related Health Damage Compensation and Prevention Association.

Environment Agency of Japan. 1998. *Commentary of Environmental Law*. Tokyo: EAJ.

Environment Agency of Japan. 1999. *Technical Manual for Air Pollution Control*. Tokyo: EAJ.

Environment Agency of Japan. 2000. *Quality of the Environment in Japan, 1998*. Tokyo: Planning Division, Global Environment Division, EAJ.

Hashimoto, Z. 1985. "Information system for environmental management", paper presented at international workshop on Environmental Management for Local and Regional Development, Nagoya, Japan, 9–13 June 1985.

Imura, H. 1993. "Air pollution control policies and the changing attitudes of the public and industry: Paradigmatic change in environmental management in Japan", in *Environmental Pollution Control: The Japanese Experience*, proceedings of the UNU International Symposium on Eco-Restructuring, 5–7 July 1993. Tokyo: UNU.

Japan Environmental Corporation. 1998. *Environmental NGO Directory*. Tokyo: JEC.

Japan Environmental Council (ed.). 2000. *The State of the Environment in Asia 1999/2000*. Tokyo: Springer-Verlag.

O'Connor, D. 1994. *Managing the Environment with Rapid Industrialization: Lessons from the East Asian Experience*. Paris: OECD.

Schreurs, A.M. 1996. "International environmental negotiations, the state, and environmental NGOs in Japan", Occasional Paper No. 14, Harrison Program on the Future Global Agenda. University of Maryland.

Sotome, M. 1997. *Japan's NGO Activities and the Public Support System*. Tokyo: Foreign Press Centre.

Ueta, K. 1994. "The role of local government in urban environmental management", *Regional Development Dialogue*, Vol. 15, No. 2, pp. 146–154.

Prescriptions for environmental governance

9

Chemical governance in East Asia

Glen Paoletto and Cindy Termorshuizen

Investments related to the chemical industry have been a vital part of economic growth in the East Asian region over the past 50 years. The chemical sector remains a major part of the world economy, and East Asian governments continue to place priority on attracting chemical corporations to their borders. US chemical exports alone reached $68 billion in 1999, which was actually low compared to the exports of 1998, due to the strong US dollar and a fiercely competitive global market (American Chemical Society 2000). Chemical corporations offer much in terms of economy, future industry, know-how, and employment. The value of the chemical industry to national and global economies and the potential risks of its products require a balancing act on the part of corporations and other stakeholders, namely government, the financial sector, labour, and consumers. This balancing act is what chemical governance is about.

Building on the work of the United Nations University (UNU), this chapter discusses some future directions of governance related to the chemical sector in the East Asian region, focusing on information systems and disclosure as a driving force for change. It examines the needs of governance pertaining to chemicals in the light of global trends, and raises options to maintain both competitive edge and consumer trust in the East Asian region over the longer term.

Economy and environment

While economic growth in the East Asian region is clearly dependent upon national policies and laws that can direct and attract desirable corporate activity, the global economy and its rapidly shifting trends impact significantly on the decisions that governments have to make (Dobson and Chia 1997; Wade 1990). Foreign direct investment (FDI) still accounts for a large part of how the East Asian region as a whole is developing, and East Asian countries have come to rely on investments from corporations based in the USA, Europe, and Japan as one of the most important means of development. To attract FDI, competition is fierce and governments go to great lengths to obtain it, in particular ensuring sources of comparatively cheap labour and tax concessions.

While corporations are able to finance development, they are also responsible for creating the vast majority of the world's pollution. Turning to chemical corporations, the world's chemical companies contribute US$1.5 trillion to the global economy (about 3.5 per cent) and hundreds of thousands of jobs. Globally, they produce about 85,000 chemicals for a multitude of products, and add about 1,000 new chemicals to that list every year. It is also interesting to note that most of these new chemicals are developed in the USA. For many of these chemicals, only time will tell whether they are environmentally benign – the usual battery of tests are generally not sufficient to provide information about their long-term behaviour in and impacts on the environment. Scientific uncertainty and the sheer volumes of information involved make answers difficult to come by. However, the devastating environmental and human health effects of chemicals such as DDT, PCBs, and dioxins have drilled home the need for attentiveness and diligence (Colborn, Dumanski, and Myers 1997).

Among these economic processes, however, lie the future-oriented needs of the environment. Already the environment is being seriously threatened at a global level, yet needs to be kept intact for our own long-term survival. The East Asian region is home to a sizeable share of the world's population, and as such the impacts of its actions go beyond the borders of any one country. It has been already noted by leading analysts that chemical markets in the developing countries of Latin America and Asia are expected to have comparatively higher growth rates over the longer term due to high population growth, improving living standards, and rapid industrialization (Standard and Poor's 2000).

When considering the chemical sector and environmental governance in East Asia, a mixed situation emerges. In this geographically wide and culturally diverse region, there are areas where sophisticated technologies are being employed in production and refining facilities, while there are other areas where antiquated and even banned chemicals continue to be produced, sold, and applied. Overall, it can be said that the legal frameworks in terms of protection for the environment and health are relatively weak in East Asia. As one can see from the earlier chapters, such legal frameworks are in place and sometimes "well drafted", but they do lack the effective enforcement capabilities needed. Obviously, problems such as a lack of an appropriate enforcement mechanism, existing corruption, and poverty make application difficult.

Increasing complexity

Factors related to the global economy and environment at the macro level are in many ways beyond the control of national governments in East Asia (Brian, Durlauf, and Lane 1997). Complexity is perhaps an increasingly important factor in evolving paradigms of governance. Whether governments desire it or not, complexity impacts directly on how a government governs.

The basic idea of complexity is as follows: all systems on earth, including economic systems, follow a basic tendency to become increasingly complex as they evolve (Peak and Frame 1994). With a system's evolution comes its ability to link with other systems and impact on those, even though – institutionally – they may seem of little relevance. Thus, with more information and progress, the paradigm comes to look like a spider's web where all the parts are linked and interlinked in terms of cause and consequence. Some have suggested that the global economy and all the systems that support it in a basic sense have by now reached this level. Globalization has been perhaps the most important compounding factor of complexity, though difficult to measure – the pace of change accelerates; new types of cooperation become needed; business and industry restructure themselves; technology develops; unknown changes occur in the natural ecosystem; information becomes freer; while population growth and ageing issues have new implications.

Complexity in a socio-economic context may be better understood by thinking on two principal fronts: a "real" increase in complexity, as economies and populations expand and the rate of technological change speeds up; and a "knowledge-induced" complexity, as previously unstudied natural and other systems become increasingly understood to be highly complex (Termorshuizen 1999).

In relating this to the chemical sector, chemicals incorporate both real and knowledge-induced aspects of complexity. In the USA alone, at least 70,000 chemicals are currently being produced (real complexity) (Mazurek 1998). We know that chemicals need to be monitored and tested for safety, but the sheer number of chemicals makes government supervision of all of them impossible. For example, the US Environmental Protection Agency (EPA) has its toxic release inventory programme requiring companies to provide information on releases and transfers of designated toxic chemicals and compounds. That programme covers less than 1 per cent of the total chemicals produced in the USA. In 1993 only 319 chemical categories required reporting by the manufacturer; but this had more than doubled, to 652, by 1995 (EPA 2000).

To put it another way, it was mentioned earlier that each year about 1,000 new chemicals are released. To test those 1,000 new chemicals more than 166 million tests are required. To test those chemicals for long-term effects, such as cancers and endocrine disruption, each of these 166 million tests would take a minimum of two years. Yet the reality is that in the USA perhaps only 500 tests are undertaken each year, and even this number is paid for reluctantly (Mitchell 1997). The result is complexity of a highly unknown nature.

Adding to the difficulty of monitoring and regulating chemical production is what is known as multiple chemical sensitivity (knowledge-induced complexity). While many chemicals might be environmentally benign under laboratory conditions, interactions with other chemicals under real conditions can modify properties and make them hazardous. Of course, testing all the 85,000 chemicals produced globally for all possible interactions with other substances is simply not possible.

Governance, governments, and complexity

What does increasing complexity mean for governance? In terms of governance, the existence of complexity factors means that it has become extremely difficult for governments to govern effectively with only the traditional approaches, and that "mixed options/approaches" become preferred choices to provide solutions.

"Good governance" is a term used by development banks to give recognition to the role of governments in the overall governance regime. In the context of a rapidly evolving paradigm, governments are faced with formidable challenges. Developing country governments – like most of those in East Asia – have a particularly difficult job. They have to find a way to govern that can promote the objectives of economic growth, stability, and employment in a rapidly changing and difficult-to-keep-up-with world marked by complexity, competition, and environmental pressures. The governments have made and continue to make decisions about which industries and sectors to support, and which to move away from (Wade 1990). Consequently, governments play a central role in how chemicals and their effects are managed, particularly in East Asia. Governments are obligated to take the first steps in safeguarding the environment and the people, both the present and future generations. The way in which a government does this is closely tied to, on the one hand, the level of economic development, and is pressured by, on the other hand, global economic trends.

Moves away from centralization

There are other factors related to institutional changes that East Asian governments must learn to deal with. Historically, centralization has always been a part of economic growth. One of the more important incentives for centralization was the promotion of economic activity and development (Aglietta 1979). To compete in the early years of capitalism, countries had to mass produce, requiring large plants, factories, and electricity providers. Also needed were large facilities (ports and harbours) and massive capital and reinvestment. In this scenario, centralized control is the more efficient means to achieve economic objectives. Efforts can be focused, and results can be visibly seen comparatively quickly. Yet today the situation appears to be changing rapidly, with implications for East Asian countries. Global economic trends and environmental realities are moving away from mass-production systems.

Production and market systems are, today, much more flexible and very fast. In economically powerful countries, "big" is no longer the market, and mass production is no longer the system. "Service" is now the market and "efficiency" is the system (Pine 1993). The "global factory", we now know, also works to eat away the foundations of sovereignty, as a country's practical significance is necessarily reduced. Consequently, protectionism is by now a policy that has become less efficient, and competition has become the norm. At the base of all these changes lies information technology and telecommunications. While not all these changes impact directly on developing countries in the region in terms of economy, where mass production is often still the norm, global economic realities both force and induce significant changes (Rhodes 1996). The numbers of corporations and activities, the often unclear delineation between public and private entities, and a host of other issues make it near impossible to govern effectively using traditional paradigms. As one example, the number of international companies in China has well surpassed 250,000 (Li 1997). This means that the way governments govern rapidly changing sectors, such as chemicals, needs continual reinvention.

Role of governments in an East Asian perspective

Despite the fact that fundamental ideas and needs are changing rapidly, governments in East Asia find it difficult to keep up. Their systems and structures are currently more suited to a mass-production and standardization-type approach than real service provision. National governments in East Asia essentially remain structured as centralized organizations, prone to top-down government based on command-and-control functions.

Most national ministries in East Asia (even in developed countries) are absolutely vertical in structure, and focus on topics or issues – industry, trade, environment, health and welfare, technology, education, and so on. Within these broad categorizations, bureaux are given broad-parameter responsibilities; within those parameters, departments are assigned responsibilities; and within those, sections are assigned more specific responsibilities, and so on. Ministries are not encouraged to take an interest in another's territory. The information tends to flow from top to bottom, at which level there is some exchange of information. While local governments in East Asia have varying structures, they too remain largely centralized and one-leader oriented.

Results of UNU research further support the view that straightforward command-and-control regulation in East Asia is not functioning as well as one might hope (Ahmad and Ali 1999). As mentioned, a centralized system can enable governments to introduce tighter standards and regulations on the behaviour of chemical corporations and individuals. While sound in principle, the serious downside of a command-control approach is the copious amounts of money and human resources needed to support monitoring and policing agencies, together with relatively non-complex scenarios. With most governments facing shrinking budgets and rising complexity, the reality therefore becomes far from the intent. In a new economic paradigm of more information and rapid changes, stubborn centralized and closed governance systems will eventually collapse from overload.

How to govern?

The key issues to focus on for governance are the seemingly opposed objectives of a vigorous economy and a more secure environment. Research conducted at the UNU provides a few important clues – three underlying fundamental requirements are noted:

- the importance of reliable information systems;
- linking these systems to innovation;
- focusing on certainty.

Information systems and governance

The advance in information technologies has effectively undermined closed economies; more open and flatter systems of information are slowly coming to the fore – more for reasons of efficiency and practicality

– and there is evidence to suggest that these same systems can be better utilized to promote good governance. Parameters will be needed, however, to define better the types of documents made public, and at exactly what point in the legislative or other process they are made public (Institute on Governance 1996).

Results and recommendations of East Asian country studies on chemical governance commissioned by the UNU repeatedly stress the importance of governments as *information managers* as opposed to decision-makers. This recommendation, based on the findings of the studies, would seem to make sense – given the trends and directions spoken of earlier. However, mechanisms for accountability, information sharing, cooperation, and access are usually not high priority for government departments in any country. For instance, the police, courts, attorneys, and counsellors all have the same goal, to maintain law and order, yet rarely talk to each other, let alone cooperate. In environment, the same paradigm tends to apply (Heaton, Repetto, and Sobin 1991).

Accountability is generally recognized as an important component of good governance. It is seen as a major motivating factor to bring about more responsive bureaucracies. Information disclosure and reporting mechanisms directly support accountability. The labour market in the EU is one example. By making the preparation of national action plans open to the scrutiny of citizens as well as other member states, the degree of political accountability for and reliability of commitments made has increased (Agence Europe 1998). In East Asia, Singapore seems advanced with its PS21 project (Public Service in the Twenty-first Century). PS21 aims to improve the efficiency and effectiveness of government, as well as accountability. A large part of this strategy revolves around the incorporation of information technologies into government (Government of Singapore 2000).

Appropriate (well-structured) information systems can do more than perhaps any other mechanism to promote accountability of government. Information systems targeting accountability can also function at the individual firm level – as an effective alternative and support to end-of-pipe-style regulation. Pollutant release and transfer registers (PRTRs) now exist in many industrialized countries, and are widely credited with significant declines in chemical waste being released into the environment by industry over the past 10 years. Internet sites set up by environmental groups explain in a very effective way (easy to understand) complex chemical information to the public, and have worked to add effectiveness to these reporting requirements (see the Environmental Defense Fund's chemical scorecard at www.scorecard.org as one good example of what is possible). Naturally, the internet is not available everywhere in East Asia, and the size of the population makes it difficult to raise awareness. Yet slowly the situation continues to change in these ways.

Basic "access to information" legislation is also worth mentioning in a discussion of information and governance. This type of legislation exists in only some of the advanced industrialized countries, and many have no such legislation as yet. National-level freedom of information laws exist in Australia, Canada, Ireland, and the USA. The European Union has a directive specifically on freedom of access to information on the environment. The UNECE Convention on Access to Information, Public Participation in Decision-making, and Access to Justice in Environmental Matters was adopted on 25 June 1998 in the Danish city of Aarhus at the Fourth Ministerial Conference in the "Environment for Europe" process, but is seen to be of global significance as the first of its kind promoting "environmental democracy". The Republic of South Africa is presently considering similar legislation, and campaigns in Great Britain and Israel continue.

There are benefits to having such laws – they can preserve citizens' right to information, which can work to provide a context in which more information can be disclosed. In practice, trade-secret provisions and policies act as an impediment to the disclosure of corporate information (Robbins 1994). However, evidence from the USA suggests that chemical companies rarely attempt to classify chemical use data as trade secrets (Ranganathan 1998).

Linking information to innovation

Corporate governance must promote innovation, competitiveness, and superior environmental performance. These are an essential part of the government's updated role that must now go beyond traditional regulation (Smart 1992). That the first two points of innovation and competitiveness are closely linked is fairly well accepted (Day 1998; Mandel 1998). Given the rapid pace of technological change in key areas of modern economies, including finance, telecommunications, and biotechnology, innovation has become much more central to the survival of firms. So information disclosure systems for the environment also need to be structured to promote innovation and efficiency. It is fairly well recognized that substandard natural resource management and waste management translate into high clean-up costs, low efficiency, negative reactions, and lost sales. At this stage of the discussion, incentives for information disclosure (taxes, threats by government, prizes, public recognition, increased sales) become a topic for further thought.

Still, in East Asia, command-and-control-style regulation, defined by heavy emphasis on emissions standards and licences, has been and continues to be the major tool used by governments to govern corporate behaviour. This is quite obvious from the chapters included in the first three sections of this volume. However, building on the information trends of late, different types of regulation as well as sets of voluntary measures have become increasingly important. It is the voluntary measures that this chapter is perhaps most concerned with, because they offer a new range of techniques that are severe enough to ensure regulation without legislation (meaning less cost for the government and to the economy), yet flexible enough to promote the innovation that a new economy demands.

The shift away from command-and-control approaches has a number of facets. On one level it has resulted from a sense of frustration at the failure of command and control to reduce environmental impacts. The situation is particularly severe in many developing countries, where capacity to implement and monitor regulations is limited. A further reason is the economywide trend toward deregulation spurred by both rapid technological change and trends in economic and political thought (Vogel 1996). Information disclosure and reporting systems go to the very heart of the matter. The increased use of disclosure requirements is particularly marked in the area of finance and banking (Nicholl 1996).

Disclosure requirements, however, are not perfect. One major flaw is their failure to take into consideration the need to allow people to understand the information disclosed, and to use it. This flaw has particularly serious consequences in East Asian developing countries.

Focusing on certainty

While we have much to learn about chemicals, there is much we already know, and effective information exchange can go a long way to reducing or eliminating many of the current problems in East Asian chemical governance.

A prudent approach to governance would naturally promote information-sharing mechanisms for what we do know about chemicals and their reactions. The recent negotiations for a persistent organic pollutants treaty highlights this approach (EPA 1998a). The US Center for Disease Control and Prevention (CDC) publishes a useful set of guides on the health effects of toxic chemicals in its *Toxicological Profiles* series (CDC 1999). Other countries and international organizations also produce similar volumes, providing the scope for increased information exchange and knowledge about the effects of the main offending chemicals. In addition, scientists are now refining the existing protocols to "screen" chemicals based on their molecular structure alone (Visva-Bharati University 1998). These computer-based procedures do not require any physical testing and can significantly (by orders of magnitude) narrow down the horizon in terms of the number of tests that need to be carried out. These are the types of systems that need to be focused on to ensure statistically consistent information required for governance decision-making.

Options for effective and improved governance

What options do governments have for effective chemical governance? Focusing on improved information disclosure systems as a means to reach a goal of improved environmental conditions at minimal cost, options for governance come within three general categories:

- regulatory mechanisms;
- voluntary mechanisms;
- market mechanisms.

In terms of governing chemical corporations, government officials need to understand the options available to them – perhaps a "mix" of these and other mechanisms is needed to reach a policy or governance solution.

In terms of practical issues, a recent survey by UNEP and SustainAbility (UNEP-SustainAbility 2000) regarding oil sector disclosure systems found the following.

- Of the 50 companies surveyed, 28 are undertaking some form of systematic social disclosure, but approaches vary widely in both form and content.
- Incentives for greater corporate social accountability include protection of licences, avoidance of risk from sensitive social issues, and staff recruitment and retention.
- Key areas of stakeholder concern have been reflected in high-profile (but relatively short-lived) campaigns, such as those focused on human rights, distribution of corporate benefits, indigenous rights, and labour conditions.
- There is a pressing need for greater convergence and coherence in the area of social reporting.

Regulatory mechanisms

Regulatory mechanisms refer to those sets of rules that are enacted in legislation, administered by governments, and backed by the threat of legal sanction or tax incentives. They vary widely between countries and over time. In making laws, governments may or may not consult the relevant industries for input. The rules themselves can be technology specific or results oriented, and may or may not rely on market incentives such as pollution charges and tradable permits. Regulatory mechanisms at the international level share many of the traits of national-level regulation. However, international law is much weaker than national law, and has not yet had a significant impact in East Asia. Still, international environmental law processes do tend to have a hidden educational and awareness-building effect.

In some countries, laws have been of critical importance for several decades in informing citizens and investors about corporate environmental performance. However, a much newer phenomenon is the specific inclusion of regulations calling for extensive information disclosure related to corporate resource efficiency and waste management. Some of the most comprehensive in this category are found in PRTR laws – that is, laws encapsulating the pollutant release and transfer register. A PRTR is a "catalogue or register of potentially harmful pollutant releases or transfers to the environment from a variety of sources" (OECD 1996). PRTRs are not necessarily regulatory measures, but in practice usually take a regulatory form. They presently exist in at least seven countries: Australia, Canada, England, France, Japan, South Korea, and the USA. Mexico has a partial PRTR covering 140 chemicals (Sissell 1998). Individual governments establish which pollutants to classify as "potentially harmful" and which sources (mainly industrial sites) will be required to report on them. PRTRs are generally made available to the public, excluding only that information which is classified as trade secrets.

The OECD argues that the information disclosed in PRTRs promotes pollution prevention for three reasons. First, it assists government authorities in setting priorities for reducing or eliminating the most harmful pollutants. Second, it encourages pollution generators to re-evaluate their production processes by showing the extent of resources lost as waste, and the moneys needed to treat that waste. Third, it instils competition among pollution generators based on the fear of being labelled as "polluters" by the general public. The USA and Canada have had PRTRs in place since 1986 and 1993 respectively, and significant reductions in pollutant releases to the environment have been documented in both countries. According to the US EPA, between 1988 and 1995 releases of designated pollutants to air decreased 46.1 per cent, to surface water 78.2 per cent, to land 42.2 per cent, and to areas underground (by underground injection) 15.6 per cent (EPA 1998b). In Canada, between 1995 and 1996 alone, on-site releases reported had declined by 14.9 per cent (Environment Canada 1996).

Industry spokespeople support the idea that these reductions are related to PRTRs. For instance, Dupont Corporation, the world's largest chemical company, claims that its position as number-one polluter on the 1988 Toxic Release Inventory (the US PRTR) led the company to take a more proactive position on emissions reduction and pollution prevention.

This was despite the fact that all its emissions were within the current legal limits. The CEO noted that "given the tenor of the times and the mood of the public, these emissions would not be tolerated indefinitely" (Smart 1992). Between 1991 and 1997, Dupont reduced its total waste generated by 46 per cent (Dupont 1998).

Strengthening PRTRs

PRTR laws target transfer and release data, but they can be significantly strengthened by shifting their focus from reporting on toxic emissions to chemical processes at the time of production. Corporations would feel pressure to reduce waste and minimize health risks at the source of the problem – where the waste or problem is produced in the first place. The current systems of chemical governance have to do a better job at showing that reductions in pollutant releases to the environment can be equated to resource efficiency innovation and higher levels of competitiveness. In other words, governance needs to show that waste saved is money made, and that innovation to reduce wastes and control risk at the source (origin) means new markets and profitability. The experience is that information needs to be presented so that links are clearly understood, and are backed by market and tax incentives to force innovation from both corporations and governments.

The result is a needed "shift" in *what* to report – from reporting on the end result of emissions to reporting on the chemicals that are used at the time of production and how they are applied. This focus would shift manufacturers' focus from waste management to process innovation. As Kevin Mills notes, "[t]o put pollution prevention first, right-to-know programs must be expanded to track toxins in manufactured products, within manufacturing processes, and in upstream activities such as material processing and transportation" (EDF 2000). In summary, chemical and corporate governance needs to do more than protect the environment; it needs to respond to and promote the economy. It needs to create jobs and markets, as well train and educate. The global economy has shifted its focus from low to high tech, which means that increasingly innovation is needed by and from corporations and governments to compete. Chemical governance needs to spur innovation by chemical corporations and governments. PRTRs offer one choice for governmental action in East Asia.

Reorienting systems

For expanded information disclosure and reporting systems (laws, policies, and programmes) to bring about increased innovation, competitiveness, and environmental performance, legal, administrative, and bureaucratic systems need to be reoriented. Reorienting in this sense essentially means a "rethinking" of the systems.

Present systems often stifle the innovation that process-oriented information disclosure could spur. India is an example in point. India is a country with a large amount of information on pollution emissions available to the public. Under law, industries in India are required to submit detailed reports identifying both the sources of pollution and any control measures in place, but tax and subsidy structures do not provide industry with any incentives to reduce pollution at the source. Polluters are not even given incentive to choose least-cost pollution abatement technologies (Murty 1995). In other words, they are required to report, but there is no priority for follow-up action. As an example, there are dozens of pesticides banned internationally that continue to be used in Indian agriculture.

As an example of reorienting systems, choices by the *consumer and market* based on efficiency ratios and ratings would seem to make more sense. A good example of data that can be used by a government or system to rank technology according to efficiency has been implemented by the UNU's Choice by CO_2 consumer choice series in a Japanese context (GEIC 1998). Such a system of governance would work like this:

- technologies are ranked according to energy and other efficiencies;
- technologies are discriminated according to ranking;
- more efficient technologies are backed by tax incentives.

Using information to rank products and technologies according to efficiency ratios, systems can be set up to promote the most efficient technologies. Once technologies are ranked (and information already exists to do this), then technologies can be discriminated on sound scientific principles, based on mathematical formulae. Provided there is political will, it is easy to foresee tax incentives being given to technologies coming within the upper echelon of efficiency ratings. The market would still be given a *choice* – less efficient technologies can still be bought, but the market would also be forced to compete with itself to bring about more efficient technologies and products at more competitive prices (Weiszacker, Lovins, and Lovins 1997).

Voluntary mechanisms

Many governments have come to realize that legislation alone is not enough to achieve the objectives demanded by public good. Until recently, though, the alternatives were considered to be relatively few. In the last 10 years, however, a new grouping of mechanisms targeting corporate and chemical governance have come about that provide more options for governments in dealing with complex issues. This shift away from command-and-control regulation has meant an increasing emphasis on voluntary mechanisms for reducing industrial and chemical emissions. There are three types of voluntary mechanisms: unilateral commitments, public voluntary programmes, and negotiated agreements (Labatt and Maclaren 1998).

To introduce the topic of voluntary mechanisms, one can look at how these mechanisms have been used without government initiative and to what effect. Environmental health and safety (EHS) reports have been a main means of voluntary information disclosure for individual firms. The number of corporations preparing these reports has increased markedly in the past five years (Davis-Walling and Batterman 1997), but the methods have not proven very successful in terms of promoting innovation. Results seem to indicate that the stated objective of the reporting (or what it is perceived to be) significantly affects the contents of the reporting. Crossley and Points (1998) argue that the bulk of investors see environmental health and safety reports as largely qualitative documents whose purpose is not related to establishing links between a company's environmental spending and its profitability. Three lessons from the EHS reporting experience can be drawn.

- Who checks? Information systems for the purposes of governing corporate activity need quality control systems to ensure that the information is comprehensive and trustworthy. Under EHS, many companies choose to report only that information which enhances their public image (Davis-Walling and Batterman 1997). In short, this practice is lying, with a resulting distortion of the facts. In addition, firms in high-impact industries, like fossil fuels and chemicals, do a significant part of their production overseas but fail to report on their overseas operations.
- Smaller players miss out. Smaller corporations tend not to make environmental (efficiency assessment) reports. This is understandable given the expense involved in compiling a report, but there are some industries where the contribution of small firms to overall pollution and innovation is significant. For example, the Environment Agency of Japan reports that dry-cleaning and hairdresser shops in Japan (usually run by individuals and small businesses) are major emitters of toxic waste (*The Economist* 1998). Where small firms are significant producers of toxic emissions, support from government will be necessary if they are to report. Yet government also needs to partner with others to bring about effective results. Universities, for example, have a real role to play to help make small and medium-sized firms more efficient, especially in developing countries (Inoguchi, Newman, and Paoletto 2000).

- Is it accurate and how do you compare? When using information for purposes of governance, the information needs to be both accurate and, most importantly, understandable by ordinary citizens to compare results so that the information has meaning. Incentives then need to be given for a change towards increased efficiency and innovation, as noted above.

Voluntary unilateral commitments

As a voluntary mechanism, unilateral commitments are undertaken by corporations either alone or as a group. However, there are serious problems with these types of arrangements. They suffer from a lack of credibility, especially due to no third-party assessment. The unilateral commitments by groups of firms under the auspices of the Japan Federation of Economic Organizations (Keidanren) is a typical example, including large chemical corporations. In this case, public and other participation is virtually non-existent or at least kept to a minimum (Keidanren 1997).

In addition to voluntary activities undertaken by individual firms, many cooperative arrangements exist among corporations, but almost always within particular sectors or industries. The chemical industry's Responsible Care Program is an example of this (Labatt and Maclaren 1998). Launched by the Canadian Chemical Producers Association (CCPA) in 1985, the programme has spread to over 40 countries and accounts for 86 per cent of world chemical production (by volume). Responsible Care targets changes in management practices rather than developing environmental performance standards. Member companies agree to improve their environmental, health, and safety performance continuously and strengthen public outreach. The programme's underlying rationale is that the improvement of the chemical industry's public image, battered by numerous accidents and scandals, is critical to its survival. In part, the programme also finds rationale because permits to expand operations are difficult to obtain when local residents and NGOs vigorously oppose them. Hostile national legislation is also less likely if public opinion is not strongly negative (Mazurek 1998).

Emissions reduction data achieved and public support by the Canadian and American chemical associations suggest that weak information disclosure systems, such as that of the US Chemical Manufacturers Association (CMA), lead to lower waste reduction and lower public support. Conversely, where there are strong, results-oriented information disclosure systems, the overall results appear to be more positive. As a result of the CMA approach, only anecdotal and case-specific data exist on the effectiveness of the programme. This compares to the Canadian Chemical Producers Association, which can point to concrete, industrywide reductions. These data are mostly encouraging – pointing to "lower costs from treatment of injuries, reduced losses of materials in spills, and improvements in regulatory compliance" (Mazurek 1998). The problem is that the data are widely mistrusted, which calls into question the objectives of the programme itself. The US Public Interest Research Group (PIRG), for instance, argues that most chemical companies in the country have not yet adopted even the principles of Responsible Care.

The experience points to a valuable lesson for East Asia: information systems need to be made reliable, and need to be trusted by the users, ultimately the people.

Negotiated agreements

Negotiated agreements result from negotiation between industry and government and/or non-governmental organizations. Those between industry and government only are by far the most common, and East Asian governments will generally deal with chemical governance in this way.

Negotiated agreements between industry and government are by no means new in East Asia. In Japan they have existed since the coming to surface of the notorious Minamata disease in the 1960s, and there are apparently presently more than 30,000 negotiated agreements (even though these are not "agreements" nor "negotiated" in the Western senses of the words) (Imura 1998). In the European Union they are officially smaller in number – in 1996 over 300 negotiated agreements were recognized by EU country governments (European Environment Agency 1997). As for whether negotiated agreements promote innovation, competitiveness, and improved environmental performance, opinions remain mixed. Some argue that in the negotiation process, collective learning takes place and innovation is encouraged. Others argue that the unambitious targets of negotiated agreements lead to less innovation than taxes or other approaches (Borkey and Leveque 1998).

While information disclosure mechanisms vary significantly across agreements, a number of common concerns exist. Questions of trust are one of these, as the negotiation process is often not so transparent. This can seriously undermine the credibility of these agreements and needs to be avoided by including transparency measures. One way of increasing transparency and ensuring that information generated remains in the public domain is to involve professionals in the negotiation process.

Public voluntary programmes

The third broad category of voluntary mechanisms is public voluntary programmes. These are non-regulatory mechanisms that are voluntary in law but, in effect, mandatory due to an overarching threat by government to introduce more severe, sanctioning, or non-friendly legislation should corporations not take action voluntarily. While voluntary mechanisms are non-regulatory, in most situations they are undertaken either to avoid regulation or with the understanding that a failure to abide by them will lead to the imposition of regulation. They are actually far from voluntary in the ordinary sense of the word. In the near future, it is likely that East Asian governments will also have to learn how to pressure major corporations to innovate and reduce wastes at the source *without* command-control legislative scenarios.

Public voluntary programmes are typically designed by governments and voluntarily joined by corporations. Many of these voluntary mechanisms have significant information disclosure components. For example, the US EPA announced a joint initiative of the EPA, the CMA, and the Environmental Defense Fund to test 2,800 chemicals for their environmental and health effects between 1999 and 2004. Under this programme, chemical manufacturers are given 13 months to volunteer their products for testing. Any chemicals not volunteered during this period will be tested under EPA direction, with the possibility of incurring less friendly or more rigorous testing. The Environmental Defense Fund is responsible for monitoring progress (third-party validation) and will also disseminate information to the public over the internet (making information understandable to citizens). This information is detailed, outlining progress on a per-chemical and per-company basis (EDF 1998). These types of initiatives are those that East Asian governments – in eventual cooperation with each other – can build upon to make governance of the chemical sector more effective and feasible.

Two other well-known programmes in this category are ISO 14001 and the European Eco-management Auditing Scheme (EMAS). The auditing process built into both of these programmes provides corporations with information on how to improve their environmental performance. By systematically checking that their operations comply with all regulations, companies can reduce the risk of heavy fines and legal battles as well as reduce insurance rates (Bruner and Burns 1998). The publication of information related to these audits can also assist governments, investors, and the public in identifying which companies have good records and which need to improve. This provides further incentive for corporations to be innovative, produce efficiently, and act responsibly, since their environmental record can link with the trust of the consumer and investor simultaneously. However, the approaches of ISO 14001 and EMAS to information disclosure have significant differences, with EMAS placing much greater emphasis on making information available to the public; ISO 14001, however, does not.

ISO 14001 has undoubtedly brought about benefits and has been a good compromise on many fronts, though it would benefit more from adopting a public statement akin to the EMAS environmental statement. Of course, given the relative newness of both ISO 14001 and EMAS, there is no empirical evidence to suggest that the fuller disclosure provided by EMAS leads to superior environmental performance than that which would occur under the ISO system. Further, there are problems with the EMAS system as it currently stands – it can be overly bureaucratic and burdensome on both government and corporations, and more efficient measures to achieve the goals within that system will be needed. Nonetheless, information disclosure from regulatory and other voluntary programmes does show a link between information disclosure and reporting systems and trust from investors and the public. This provides an incentive at least for the "visible" chemical corporations to innovate and reduce waste.

The ISO 14001 system has contributed positively to environmental information disclosure on a global scale. It has been particularly useful to many developing countries in providing a blueprint for national corporate disclosure rules. The Mexican Environment Ministry has added a voluntary ISO-type system to its regulatory framework, and both the Philippines and Indonesia plan to incorporate elements of ISO 14001 into new programmes for corporate environmental information disclosure (World Bank 1998).

Market mechanisms

Market mechanisms are the third in the series of mechanisms that work to govern the behaviour of chemical corporations. This chapter is more concerned with information-related aspects of governance, and therefore will not deal with these mechanisms in any great depth. Nevertheless, information systems are usually tied to all market mechanisms, directly or indirectly. Stated simply, the purpose of market mechanisms is to discourage production and discharge of toxic pollutants; encourage industry and consumers to make environmentally positive choices; and provide funding to implement other regulatory measures associated with environmental protection.

Market mechanisms that have been applied and give options for governance in East Asia include:

- **product taxes** – to ensure that the market price of products reflects the costs that they impose on the environment, particularly during use and disposal;
- **an emission charge system** – for industrial emissions, based on the polluter pays principle;
- **general requirements** – that manufacturers, distributors, and retailers take direct responsibility for disposal of products containing toxics;

- **a deposit/refund system** – to ensure the return of products containing toxic materials;
- **civil liability** – a significant expansion of the civil liability of toxic polluters;
- **mandatory insurance** – and/or security requirements for polluters, sufficient to compensate for potential "worst-case scenario" pollution damages;
- **a government purchasing policy** – that gives routine preference to products produced with clean technology (discussed to some extent in reorienting systems, above);
- **a review of all government subsidy programmes** – for elimination of non-environmental subsidies.

Banks, insurance companies, and investors play a crucial role in how an economy is shaped and what corporations do, including with regard to the environment. Without the confidence of the market, money stops flowing in and begins to flow out. It was this situation of a lack of investor confidence that essentially underscored the Asian economic crisis. The preferred explanation for the crisis combines elements of closed systems (corruption, antiquated systems), under-regulation, and reckless borrowing by corporations. The result is seen to have brought high rates of indebtedness and badly performing loans. This brought about a crisis of confidence among creditors, followed by panic and massive outflows of funds from the region.

In the market, third-party rating and auditing agencies are increasingly playing a role in providing the targeted information, and work to provide the "trust" that the system demands. While the link is taking time to emerge, poor corporate environmental performance is linking more with poor management and financial performance. For banks and insurance corporations, it is important that a corporation maintain its assets and image, particularly publicly quoted corporations. Environmental clean-up obligations reduce a firm's equity and possibly future markets. Property used as collateral for a loan may decline in value if contaminated (Labatt and Maclaren 1998). Such firm-specific issues matter to insurance companies too, which have to insure the assets, though for those firms the global environmental problems are cause for increasing concern in economic and health terms. In the USA, some 500 sites have been cleaned up so far over a period of 18 years at a cost of some $30 billion. Legal fees have consumed about one-third of this amount. Clean-up at the five-hundredth site was completed in December 1997. Clean-up costs have been averaging $25 million per site according to the US EPA. It has been estimated that the total aggregate pollution insurance coverage in the USA in 1991 was about $3.1 billion (Insurance Information Institute 2000). Data on East Asia in this regard are scarce and still emerging, but the number of sites requiring clean-up in Japan alone can be estimated at well over 10,000 (Environment Agency of Japan 2000). Using the US experience as a basis, clean-up costs for these and other sites could well be estimated at over $250 billion. These types of costs require strategy to diffuse problems before they occur to the extent possible.

The combined interest of banks, insurers, and investors is called market mechanisms, and is a powerful driving force behind the trend toward greater corporate information disclosure. As part of this trend, environmental rating agencies have an important role to play as disinterested professionals who can assess desired information. The bulk of the work to date of these agencies revolves around analysing corporate environmental risks and returns, and the focus of much of the rating activity tends to be on environmental liabilities rather than assets (Campanale 1994).

Openness on environmental performance appears to play a significant part in promoting innovation and improving environmental performance in the long run (Dupont 1998). Numerous studies of changes in stock prices after negative environmental media reports and PRTR filings show that those corporations that subsequently lose the most money also innovate the most to improve performance. While the process can be a rather painful one, information disclosure and reporting systems can lead the financial community to react, which makes corporations become innovative for less waste and risk, better public image, and higher share values.

Information packaging and the internet

If PRTRs and other information systems are to live up to their potential, then citizens must be able to access and use them. Public scrutiny can be a powerful incentive for firms and governments to innovate and improve environmental and safety records. This is particularly so for the larger corporations which are more likely to be concerned with public image and opinion, and it is those corporations that can work with subcontractors to stipulate standards for the economy and environment. Conversely, local citizens are unlikely to pressure firms if they cannot access or do not understand the data. Countries like the USA, with strong environmental groups and widespread internet access, have an enormous head start in this area. Deserving special mention are the Environmental Defense Fund's chemical scorecard (www.scorecard.org) and Friends of the Earth UK covering England and Wales (www.foe.co.uk).

For countries with weaker environmental groups and less access to information, regulation may be more effective until infrastructure becomes such that the systems open up. The longer this process takes, and the longer countries like the USA are able to innovate, the wider the gap between the rich and poor becomes. Again, the link between information, innovation, and the economy is made apparent.

To sustain economic growth in the long term, East Asian countries will have to strive for socio-economic environments that can support and promote these types of systems. In this way, governance for environment and future economic performance can be strongly linked. The internet is a powerful and understated tool for systemic information reporting and disclosure programmes. In the area of chemicals, the relatively low-cost information dissemination option that the internet can offer makes it useful and even indispensable as a link between testing facilities, government regulators, NGOs, and the public. The internet can also be used for providing training for employees, agricultural extension workers, and health professionals. Leading "e-learning" (internet-based learning) software and system providers have had a considerable degree of success in applying e-learning systems to the chemical sector. There is no reason why this cannot be extended further to governments and other related sectors.

The problem in some countries is the lack of widespread access to affordable telephone lines and infrastructure. In 1993 a review of telecommunications development in the Asia Pacific region (broader than the East Asian region) found that "three-quarters of all households did not have access to a telephone" (Mansell and Wehn 1998). Meanwhile, at the global level, fibre-optic cables have transformed the economics of international calls. A one-minute call on a transatlantic cable 40 years ago cost $2.44; in 1996 the same time cost slightly more than one cent (*The Economist* 1997). The internet grew more rapidly in 1999 than during the previous four years, expanding by 67 per cent. The year 1999 marked a turning point for some developing nations, which made their debut on the list of countries with the most internet users. China (6.3 million users) and South Korea (5.7 million) overtook several European nations to join the top-10 list. Net access in developing countries grew 93 per cent in 1999, outstripping the growth rate for the internet as a whole (WWI 2000).

Government online

There is obvious potential in the internet to promote improved systematic information reporting and disclosure systems. One obstacle for the time being is the level of awareness of many developing country government offices of the value of the internet and the purposes it can serve. Centrally controlling the internet in East Asia and treating it as a tangible, tradable commodity rather than the "web" it really is may work to hinder the innovation needed in the region.

Sufficient infrastructure exists in many countries for better governance. Putting services online such as information (e.g. local regulations) and instructions, licence renewals, payment of fines, and polling can serve citizens better. Issuing multi-cards that incorporate a driver's licence, vehicle registration, library cards, debit and charge cards, welfare assistance, and insurance verification is another way of streamlining offices and services. In Singapore, the civil service computerization programme has also meant profits – S$2.7 have been generated for every dollar spent, or more than S$100 million every year.

Information exchange

Related to internet usage and the media are issues of information sharing. As already mentioned, the joint initiative of the US EPA, the Chemical Manufacturers Association, and the Environmental Defense Fund to test 2,800 chemicals includes incentives to share information internationally. This should mean that testing is *not* extensively required for those chemicals already tested. But as UNU research indicates, knowledge of the benefits that this and other programmes like it could bring with appropriate partnership is not widespread. Access to the information and the systems to recognize information as a powerful force in environmental governance are currently not present in East Asia.

In the chemical industry, given the sheer number of tests and resources involved, developing effective mechanisms for discovering chemical risks will prove a difficult challenge and will probably come to rely on enhanced information sharing. Some parts of the world, like the European Union, have regulations requiring specified information and risk assessments for all new chemicals as well as programmes to test existing substances. However, even where programmes exist, there is considerable scientific uncertainty and progress is often slow. Efforts are being made to redesign risk assessment programmes to make them more efficient and more accessible to the public. The enormous amount of effort and information required to assess chemical risk necessitates regional and international cooperation, and obviously calls for the need to share information.

The way ahead

This chapter has largely focused on the role information can play in environmental governance, particularly when dealing with the chemical industry. Appropriate implementation of this governance regime demands information as well as innovation at the source of emissions and potential problems. Existing institutional infrastructure and prior experience are at hand in some East Asian countries to diffuse further the role of information within governance, while others still need considerable progress. The East Asian perspective on this needs to be reoriented such that reliable information systems are available and environmentally positive market mechanisms are championed. It cannot be overemphasized that well-planned information systems in traditionally regulated environments have been shown to facilitate higher standards of practice. It is equally important to remember that economic trends, both national and global, have and will continue to hold priority over the environment in the near future. This is particularly true for the still-growing economies of East Asia. Therefore, effective use of market and voluntary mechanisms can pay rich environmental dividends and help improve the performance of the governance regime.

REFERENCES

Agence Europe. 1998. Europe Documents, No. 2090/2091, 10 June. Brussels: Agence Europe.

Aglietta, M. 1979. *A Theory of Capitalist Regulation*. London: New Left Books.

Ahmad, A. R. and H. Ali. 1999. "Contaminants in drinking water (Malaysia)", paper commissioned by the United Nations University, Tokyo.

American Chemical Society. 2000. *Facts and Figures For The Chemical Industry*, Vol. 78, No. 26, 26 June.

Brian, A. W., S. N. Durlauf, and D. A. Lane. (eds). 1997. *The Economy as an Evolving Complex System II*. New York: Perseus Publishing.

Borkey, P. and F. Leveque. 1998. "Voluntary approaches for environmental protection in the European Union", paper presented to the Workshop on the Use of Voluntary Approaches in Environmental Policy, OECD, Paris, 1–2 July.

Bruner, L. J., and M. J. Burns. 1998. "The ISO 14000 series: Business-friendly environmentalism", *Journal of Environmental Regulation and Permitting*, Spring, pp. 17–19.

Campanale, M. A. 1994. "Green investment: Incentives for disclosure", *Review of European Comparative and International Environmental Law*, Vol. 3, No. 1, pp. 43–48.

CDC. 1999. *Toxicological Profiles*. Washington, DC: Agency for Toxic Substances and Disease Registry.

Colborn, T., D. Dumanski, and J. P. Myers. 1997. *Our Stolen Future*. New York: Penguin Books.

Crossley, R. and J. Points. 1998. *Investing in Tomorrow's Forests*. Gland: WWF.

Davis-Walling, P. and S. A. Batterman. 1997. "Environmental reporting by the Fortune 50 firms," *Environmental Management*, Vol. 21, No. 6, pp. 865–875.

Day, R. M. 1998. *Beyond Eco-Efficiency: Sustainability as a Driver for Innovation.* Washington, DC: World Resources Institute.

Dobson, W. and Chia, S. Y. (eds). 1997. *Multinationals and East Asian Integration.* Ottawa: International Development Research Centre.

Dupont. 1998. *Safety, Health and the Environment: 1997 Progress Data.* Wilmington: Dupont Corporation.

EDF. 1998. "Industry to test 2,800 major chemicals for health, environmental effects", EDF press release, 9 October.

EDF. 2000. *Annual Report 1999.* New York: Environmental Defense Fund's Pollution Prevention Alliance, www.edf.org.

Environment Agency of Japan. 2000. *Japan State of Environment.* Tokyo: Environment Agency of Japan.

Environment Canada. 1996. *1996 NPRI Summary Report Executive Summary.* Ottawa: Environment Canada.

EPA. 1998a. *New Protocol on Persistent Organic Pollutants.* Washington, DC: EPA.

EPA. 1998b. *EPA's Toxic Release Inventory.* Washington, DC: EPA.

EPA. 2000. *1998 Toxics Release Inventory Context Information.* Washington, DC: EPA.

European Environment Agency. 1997. "Environmental agreements, environmental effectiveness", *Environmental Issues Series*, Vol. 1, No. 3.

GEIC. 1998. *CHOCO$_2$ '97 Study – Choice by CO$_2$.* Tokyo: UNU-EAJ-GEIC.

Government of Singapore. 2000. *Public Service in the 21st Century.* Singapore: Government of Singapore, www.gov.sg.

Heaton, G., R. Repetto, and R. Sobin. 1991. *Transforming Technology: An Agenda for Environmentally Sustainable Growth in the 21st Century.* Washington, DC: World Resources Institute.

Imura, H. 1998. "The use of unilateral agreements in Japan: Voluntary action plans of industries against global warming", paper presented at the Workshop on the Use of Voluntary Approaches in Environmental Policy, OECD, Paris, 1–2 July.

Inoguchi, T., E. Newman, and G. Paoletto. 2000. *Cities and the Environment.* Tokyo: UNU Press.

Institute on Governance. 1996. *Information and Communications Technologies (ICTs) and Governance: Linkages and Challenges.* Ottawa: International Development Research Centre.

Insurance Information Institute. 2000. *Environmental Pollution.* New York: Insurance Information Institute.

Keidanren. 1997. *Japan's Industry-wide Voluntary Environmental Action Plans.* Tokyo: Japan Federation of Economic Organizations.

Labatt, S. and V. W. Maclaren. 1998. "Voluntary corporate environmental initiatives: A typology and preliminary investigation", *Environment and Planning C: Government and Policy*, Vol. 16.

Li, C. 1997. *China: The Consumer Revolution.* New York: John Wiley & Sons, p. ix.

Mandel, M. J. 1998. "You ain't seen nothing yet", *Business Week*, 31 August, pp. 60–63.

Mansell, R. and U. Wehn (eds). 1998. *Knowledge Societies: Information Technology for Sustainable Development*. London: United Nations Commission on Science and Technology for Development/Oxford University Press.

Mazurek, J. 1998. "The use of unilateral agreements in the United States: The Responsible Care initiative", paper presented at the Workshop on the Use of Voluntary Approaches in Environmental Policy, OECD, Paris, 1–2 July.

Mitchell, J. D. 1997. "Nowhere to hide – The global spread of high-risk synthetic chemicals", *World Watch*, Vol. 10, No. 2, Washington, DC: World Watch Institute.

Murty, M. N. 1995. "Environmental regulation in the developing world: The case of India", *Review of European, Comparative and International Environmental Law*, Vol. 4, No. 3. pp. 330–337.

Nicholl, P. 1996. *Market Based Regulation of Banks: A New System of Disclosure and Incentives in New Zealand*, Public Policy for the Private Sector, Note No. 94. Washington, DC: World Bank.

OECD. 1996. *Pollution Prevention and Control Extended Producer Responsibility in the OECD Area Phase 1 Report: Legal and Administrative Approaches in Member Countries and Policy Options for EPR Programmes*. Paris: OECD, OCDE/GD(96)48, p. 15.

Peak, D. and M. Frame. 1994. *Chaos Under Control: The Art and Science of Complexity*. New York: W. H. Freeman and Company.

Pine, B. J. II. 1993. *Mass Customization: The New Frontier in Business Competition*. Boston, MA: Harvard Business School Press.

Ranganathan, J. 1998. "Sustainability rulers: Measuring corporate environmental and social performance", *Sustainable Enterprise Perspectives*, May.

Rhodes, R. A. W. 1996. "The new governance: Governing without government", *Political Studies*, Vol. XLIV. pp. 652–657.

Robbins, D. H. 1994. "Doing business in the sunshine: Public access to environmental information in the United States", *Review of European, Comparative and International Environmental Law*, Vol. 3 No. 1, pp. 26–35.

Sissell, K. 1998. "Mexico starts up program", *Chemical Week*, Vol. 160, No. 20.

Smart, B. (ed.). 1992. *Beyond Compliance: A New Industry View of the Environment*. Washington, DC: World Resources Institute.

Standard and Poor's. 2000. *Standard and Poor's Industry Surveys*. New York: McGraw-Hill.

Termorshuizen, C. 1999. "The role of information disclosure in corporate governance", paper commissioned by the Institute for Global Environmental Strategies, Hayama, Japan.

The Economist. 1997. "Telecommunications survey", *The Economist*, 13 September.

The Economist. 1998. "Long-term sickness?", *The Economist*, 3 October, pp. 91–93.

UNEP-SustainAbility. 2000. *The Oil Sector Report: A Review of Environmental Disclosure in the Oil Industry*. New York: SustainAbility Engaging Stakeholders Program.

Visva-Bharati University. 1998. *Proceedings of the First Indo-US Workshop on Mathematical Chemistry: With Applications in Molecular Design and Hazard Assessment of Chemicals*. West Bengal: Visva-Bharati University.

Vogel, S. K. 1996. *Freer Markets, More Rules: Regulatory Reform in Advanced Industrial Countries*. Ithaca: Cornell University Press.
Wade, R. 1990. *Governing the Market: Economic Theory and the Role of Government in East Asian Industrialization*. Princeton, NJ: Princeton University Press.
Weiszacker, E. von, A. B. Lovins, and L. H. Lovins. 1997. *Factor Four: Doubling Wealth, Halving Resource Use*. London: Earthscan Publications.
World Bank. 1998. "ISO 14000 standards for environmental management", *New Ideas in Pollution Regulation*, Appendix 2, 23 September.
WWI. 2000. *Vital Signs*. Washington, DC: World Watch Institute.

10

The future of environmental governance in East Asia

Zafar Adeel and Naori Nakamoto

It is clear that environmental governance in an East Asian perspective has unique characteristics. The preceding chapters provide ample justification of this uniqueness when compared to the environmental governance perspectives in the North; the detailed description by Paoletto and Termorshuizen in this volume provides a good contrast between the two. This comparison even extends to other parts of the developing world such as Latin America or Africa. This uniqueness has deep roots in the history of the governance structures within these countries, particularly during the twentieth century. Indeed, it is closely linked to the political history of these countries, which typically includes colonial occupation and a tortuous path for emerging democratic processes.

The rapid industrial development in East Asia during the past two or three decades has also left a strong mark on the environment and the governance structures. This development has typically been without due consideration to environmental or equity issues. As we have observed in some cases presented earlier, this has resulted in devastating effects in key areas of the environment.

The main thesis of this book is that the governance structures and institutions in East Asia should be designed by carefully considering the local perspectives. The information provided in this book may be deemed a starting point in that direction. In this context, this chapter provides the relevant information on typical governance institutions in East Asia. Particular focus is on identifying patterns of these governance structures.

The assessment of these structures brings out their inherent weaknesses and shortcomings. There are numerous constraints, both socio-economic and political, that hinder the effective implementation of the existing environmental governance regime.

As a positive sign, there are a few cases that may be cited as success stories. There is a need to identify what elements are essential for these successes and how those can be replicated on a regional basis and with a broader horizon of environmental issues. The concluding section of this chapter provides a set of recommendations based on the information presented in this book.

Governmental institutions for environmental governance

Given the cross-cutting nature of most environmental issues, it is indeed difficult to identify all governmental institutions that may impact on environmental governance. Nevertheless, in the context of the case studies provided in this book, a clear pattern of relevant governmental institutions appears. Those that play a predominant role in environmental policy formulation can be aggregated into a list; the names listed here are generic and the reader is directed towards the individual chapters for a more detailed national outlook for each of the five countries. It is also worthwhile to take a closer look at the "typical" areas of influence of each one of these ministries. Such a listing has been compiled in Table 10.1 by aggregating the respective descriptions in Chapters 2 to 8.

- Ministry or Agency of Environment
- Ministry of Science and Technology
- Ministry of Agriculture
- Ministry of Trade, Industry, and Energy
- Ministry of Construction (and Transportation)
- Ministry of Labour
- Ministry of Forestry
- Ministry of Finance
- Ministry of Health.

Based on the existing procedures and legislation, correlations or overlaps can be identified amongst these institutions. These overlaps can be viewed as synergistic in enhancing governance mechanisms. More often than not these overlaps are potential conflicts of interests between these ministries. Also, ownership or "turf possession" by one agency can lead to friction with others when the issue is clearly multidisciplinary. To understand the nature of these overlaps in the abstract sense, a schematic in Figure 10.1 attempts to demonstrate the stronger linkages when it comes to environmental governance. Ministries involved in environmental gov-

Table 10.1 Areas of influence for agencies involved in environmental governance

Institution involved	Relevant areas of influence
Ministry of Environment	Development of environmental policy and planning; environmental monitoring, research, tests, and assessment of air pollution; industrial development, import/export of hazardous chemicals, emissions requirements; toxic wastes management, monitoring and assessing violation of environmental regulations; regulations for transportation, handling, disposal, and treatment of nuclear and radioactive industrial wastes.
Ministry of Trade, Industry, and Energy	Import and export of toxic substances and import restriction on industrial wastes; allocation and management of industrial sites; supply of energy resources like petroleum and coal; research and development on new and alternative energy sources.
Ministry of Science and Technology	Technology transfer; industrial development.
Ministry of Agriculture	Monitoring agricultural pollution; management of irrigation; agricultural land reform; promotion of rural development.
Ministry of Interior	Local and provincial water management and works authorities.
Ministry of Health	Monitoring of microbial contamination; environmental conditions in workplaces.
Ministry of Construction and Transportation	Development of new construction projects; approval and performance testing of motor vehicles.
Ministry of Labour	Countermeasures against occupational diseases and improvement of working conditions.
Ministry of Forestry	Protection of forests and monitoring of forest destruction activities.
Ministry of Finance	Collection of levy for environmental pollution.

ernance revolve around two focal regimes or sectors: environment and trade-industry-energy (TIE). The former deals with environmental issues and problems directly, whereas the latter is a major contributor to environmental stresses.

A number of ministries have more direct and stronger links with the environment ministry or agency, such as science and technology, finance,

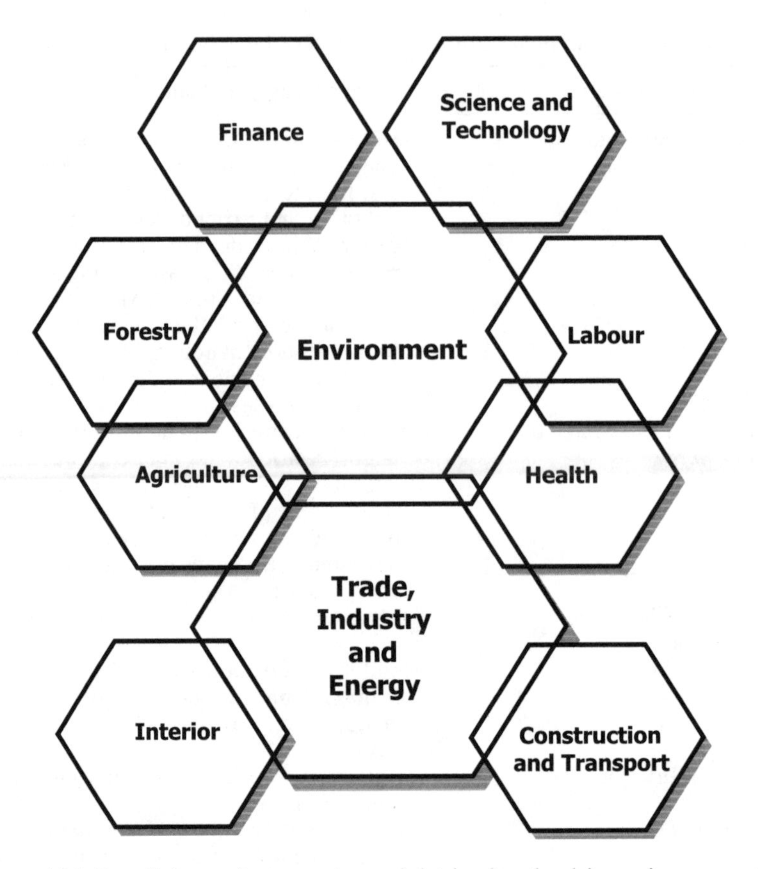

Figure 10.1 Interlinkages between key ministries involved in various aspects of environmental governance

forestry, and labour. Science and technology drives our understanding of environmental issues and problems as well as coming up with solutions to these. In the East Asian context, just as in many other regions, the ministry of finance has the final say on financial resources that are available to the environment sector. The actions and policies undertaken by the ministries of forestry and labour are usually critical to utilization of natural resources and biodiversity conservation. Ministries of health and agriculture have a comparable influence on both environment and TIE sectors. Similarly, ministries of interior as well as construction and transport typically influence the TIE sector strongly. Other overlaps, such as between forestry and agriculture and labour and health, are also important in the environmental context for obvious reasons. There are indeed other linkages, such as that between agencies dealing with labour and trade and industry. Those linkages are, however, not considered strongly relevant to environmental governance.

It is important to consider the roles of governmental institutions presented in Table 10.1 and their potential overlaps as shown in Figure 10.1. It is quite obvious that institutions dealing directly with the environment and TIE nexus predominate in the environmental governance picture. This is also quite logical in the context of the earlier discussion in this volume of the emphasis placed by East Asian countries on industrial development. The relative importance of environmental governance in national politics can also be gauged by considering the influence wielded by environmental agencies. Some countries have created a separate ministry for environmental management, while in others environment cells or divisions have been created within an existing ministry.

It may be argued that the creation of high-level apex agencies and the concentration of authority for environmental issues leads to reducing inter-ministry conflicts of interests and often enables a critical and unbiased evaluation of the performance of ministries. These agencies also provide guidance to agencies overseeing other sectors and coordinated implementation of national environmental strategies. Such an oversight is critical in linking environmental issues to national sustainable development efforts. Overall, the institutional integration of environment into economic decision-making has seen considerable progress in East Asia (UN ESCAP 2000). For example, the National Economic and Social Development Board (NESDB) in Thailand is a crucial macroeconomic institution that takes a long-term view of development issues; collates, assesses, and prioritizes the country's public investment programmes; and coordinates and monitors implementation through various ministries and local agencies. As another example, the State Environmental Protection Administration in China is the apex body monitoring and coordinating activities amongst agencies and ministries; this is achieved through four-tier networking at the national, provincial, local, and sectoral levels (UN ESCAP 2000).

Still there is a considerable room for improvement in governmental institutions dealing with environmental governance. Basic problems, like understaffing, lack of staff training, and conflicts of interests with other ministries or agencies, obstruct effective integration and implementation. Additionally, the budgetary allocation for environmental ministries or agencies has suffered during the past few years due to the regional economic crisis. Some countries in the East Asian region have reduced the budget for the environment, although there are clear differences in the precise nature of these cuts. In Thailand, for example, the budget has been reduced by about 20 per cent, particularly that earmarked for pollution reduction and energy conservation.

Constraints for environmental governance

The case studies discussed in the book describe a complex system of legislative and policy tools in place throughout the East Asian region. In addition to the complex picture created by the multitude of government agencies and potential overlaps in their roles (see Figure 10.1), the situation is further muddled up by this myriad of legislation. The interpretation and implementation of these environmental regulations is still undergoing development in the East Asian region. Compliance with environmental regulations typically remains low, essentially because many constraints are interlinked. The constraints and problems in environmental governance can be viewed from two angles: weaknesses and failures in policies adapted by the governments, and inherent weaknesses and problems in institutions relied on for implementation of policies. The major weaknesses and failures in policy formulation and implementation that have been identified in this book are summarized below.

- In the broadest sense, governments and policy-makers fail to see the correlation between environmental conservation/protection and economic/industrial growth. This leads to a lack of sufficient political will to undertake environment-friendly policy development. National development policies need to be closely aligned to sustainable development and environmental conservation – something that does not typically occur in the five countries studied.
- There is an overall lack of coordination between various ministries and agencies dealing with environmental governance. The creation of apex environmental agencies, as in Japan and Korea, has helped alleviate this problem to a limited extent.
- There is a general lack of a comprehensive national environmental policy coupled to an effective enforcement regime. The historic development of environmental legislation clearly indicates a piecemeal approach to the formulation of environmental policies rather than broad-sweeping policies. The Nature Conservation Law, enacted in Japan in 1972, is perhaps an exception to this.
- Lack of funding may be a key problem in East Asian countries, with perhaps the exception of Japan. Because of a relatively low political priority for environmental issues, funds comparable to other major sectors are not available. During the late 1990s, the economic crisis in the region further exacerbated the situation regarding the funding available for environmental issues.
- There is an inherent vagueness of standards applied in many environmental laws and regulations. This, in part, is due to the fact that these East Asian laws and regulations have been "borrowed" from developed countries and may not be appropriate for local circumstances.

- The policy-makers often fail to provide sufficient enforcement authority to the environmental institutions, resulting in weak and erratic enforcement of environmental regulations. Typically, the degree of actual compliance and enforcement varies from region to region and depends on the local enforcement officials.

The role of the international governance institutions in East Asia also demands our attention. However, the focus must be on the ineffectiveness of these institutions in protecting and preserving the environment. Three major reasons may be cited for this ineffectiveness. First, the countries in this region have identified their priorities such that the development of industry and commerce is given prime importance, whereas societal and environmental perspectives have been largely overlooked. In this context, several international and regional conventions focusing on various aspects of the environment have been ratified by the national governments; however, their implementation and action remain unclear at best. Second, a strong perception exists in these countries that these environmental issues are primarily a "Western" ethic being unduly imposed on developing countries and are perhaps an instrument to make them scapegoats for the "sins" of developed nations. This is a viewpoint that has become the stumbling block for international treaties since the Earth Summit in 1992 (Hempel 1996). Third, implementation of the international conventions and treaties has implications for the economies of these countries. Typically, these countries have looked towards the developed world for the necessary financial resources for implementation. These finances have often been quite difficult to come by – leading to the vicious circle of ill-conceived development programmes and degradation of the environment. The consequence of these three factors has been the superficial and ineffective implementation of the international environmental governance regime.

Role of industries in environmental conservation and governance

Industries play a vital role in environmental governance. This role may be construed to have two facets. Firstly, industries typically have political lobbies and political groups that can influence the formulation of environmental legislation in the first place. Secondly, the effectiveness of environmental governance is directly dependent on how industries co-operate with implementation institutions. Often the implementation of environmental regulations requires considerable investment by industries into waste management and treatment technologies. Despite the fact that short-term profitability may be adversely impacted by such adaptation, long-term benefits have been clearly demonstrated (Hempel 1996).

The notion of industrial growth before environmental conservation may define the general paradigm adopted by the East Asian nations. However, such a notion may be too broad, and we should look closely at the various key industries on a sector-by-sector basis and gauge the relative impact on environmental conservation. A summary of such environmental impacts is presented in Table 10.2; the regional growth of these industries during the past two decades is also presented to provide a perspective for quantitative evaluation. In a general sense, most of the industries have seen significant growth over the past two decades, with the exception of leather and fur products.

Observing the environmental impacts of various industries, some commonalities can be identified across the board. These include atmospheric pollution through various gaseous and particulate contamination, and waste management problems, particularly those dealing with hazardous wastes. The problems can often be traced to industrial practices that may be outdated anyway. As an example, various industries burning coal for their operation contribute from 57 to 75 per cent of atmospheric pollution attributed to coal burning, the remaining percentage is contributed by energy generation (Satterthwaite, Hardoy, and Mitlin 1992). Toxic and/or hazardous wastes from industry are generally well recognized and documented, as presented in Table 10.2. Some of the worst polluters include chemical manufacturing industries, fabricated metal, petroleum industries, and pulp and paper industries. These toxic wastes are often not amenable to more traditional treatment and disposal approaches used for municipal wastes. This means that their impacts on environmental ecosystems are typically prolonged. This persistence in the environment coupled with the huge volumes produced presents a somewhat bleak picture. As an example, China produces more than 50 million tonnes of hazardous wastes per annum (UN ESCAP 2000).

Many toxic wastes are either dumped in rivers and water bodies or on land sites with few safeguards. Reports of catastrophes resulting from improper handling of toxic wastes are frequent. A most salient landmark example is the Minamata disease in Japan, caused by improper disposal of mercury-contaminated wastes (Mitchell 1996). In that particular example, the denial by the industry in question and local authorities resulted in a long-drawn-out misery for the inhabitants of Minamata over a period of 50 years. Other similar disasters, like the Union Carbide accident in Bhopal, India, provide the impetus for developing governance institutions for managing the impact of industries on the environment.

Another "industry" not explicitly included in Table 10.2 is the agricultural sector. The first section of this book provides a clear picture of key

Table 10.2 Environmental impacts caused by various industrial sectors

Sector*	Environmental impacts
Agrochemicals	Contamination of surface water and groundwater resources during production and application; human health problems caused by accidental and chronic exposure.
Wood products *1980–1990: 6.9%* *1990–1997: 3.4%*	Sawdust discharged into rivers causing water pollution and sedimentation; indiscriminate disposal of wood waste; inefficient burning of wood waste and unapproved incineration.
Paper and pulp *1980–1990: 6.5%* *1990–1997: 7.7%*	Emissions of suspended solids, organic matter, chlorinated organic substances, toxins; emissions of SO_2, NO_x, CH_4, CO_2, CO, hydrogen sulphide, mercaptans, chlorine compounds, and dioxins.
Leather and fur products *1980–1990: 0.6%* *1990–1997: 3.2%*	Emissions including leather dust, hydrogen sulphide, CO_2, chromium compounds; water effluents from the many toxic solutions used, containing suspended solids, sulphates, chromium.
Petroleum refineries *1980–1990: 5.6%* *1990–1997: 5.3%*	Emission of SO_2, NO_x, hydrogen sulphide, HCs, benzene, CO, CO_2, particulate matter, PAHs, mercaptans, toxic organic compounds, odours; risk of explosions and fires; hazardous waste, including sludges from effluent treatment and spent catalysts.
Metals products *1980–1990: 6.6%* *1990–1997: 8.0%*	Emission of dust from extraction, storage, and transport of ore and concentrate; emission of metals (e.g. mercury) from drying of ore concentrate; contamination of surface water and groundwater by highly acidic mine water containing toxic metals (e.g. arsenic, lead, cadmium); contamination by chemicals used in metal extraction (e.g. cyanide).
Iron and steel *1980–1990: 11.4%* *1990–1997: 10.0%*	Emissions of SO_2, NO_x, CH_4, CO_2, CO, hydrogen sulphide, aromatic hydrocarbons, lead, arsenic, cadmium, chromium, copper, mercury, nickel, selenium, zinc, organic compounds, PCDDs/PCDFs, PCBs, dust, particulate matter, hydrocarbons, acid mists; risks of explosions and fires; production of wastes like slag, sludges, oil and grease residues, hydrocarbons, salts, sulphur compounds, heavy metals that lead to soil contamination; waste disposal problems.

Table 10.2 (cont.)

Sector*	Environmental impacts
Cement, glass, ceramics *1980–1990: 3.8%* *1990–1997: 9.1%*	Emission of dust, SO_2, NO_x, chromium, lead, CO, vanadium, hydrofluoric acid, soda ash, potash, silica, fluorine compounds; water contamination by oils and heavy metals; soil contamination with metals and waste disposal problems.

Source: UNIDO (2000)
* Growth in the East Asian region is shown in *italics*.

environmental issues pertinent to agriculture – the most obvious ones relate to agrochemicals and pest control chemicals. It is understandable that the rapidly growing population in this part of the world has placed an increasing demand on food and natural resources. At the same time, this growth has resulted in environmental impacts which threaten the environment and these very resources. Very significant quantities of pollution are generated as a result of agricultural activities; an overview of the amounts generated in the four countries studied in this book is provided in Table 10.3. It is quite clear that these large volumes of wastes have serious impact on the quality of water in inland and coastal waters.

A most poignant example of the challenge faced in East Asia in maintaining food security and minimizing environmental impacts can be seen in the case of rice production. As Abdullah and Sinnakkanu (Chapter 2) point out, rice is indeed the most important crop in this region. A revolution in rice production started in Asia during the 1970s and continued through the 1980s. During this period tremendous achievements in rice production were made, in which an average annual increase of more than 2 per cent was recorded in the years from 1978 to 1989 in East Asia alone (UN ESCAP 2000). To a large extent these achievements were a reflection of the desire of these nations to have food security to the point of being self-sufficient.

Table 10.3 Agricultural waste generated annually (million tonnes)

Country	Agricultural wastes	Crop residue	Total
China	255	587	842
Korea	15	10	25
Malaysia	12	30	42
Thailand	25	47	72

Source: UN ESCAP (1997)

The driving forces behind the green revolution were the technologies of chemical fertilizers for soil enrichment and pesticides and herbicides for crop management. Coupled with high-yielding rice varieties and increased land for rice production, remarkable progress was achieved and the Asian nations were able to keep up with their rapidly growing population. However, there is a price to be paid for this advanced technology and food security. The use of chemical fertilizers, pesticides, and herbicides was critical for the short-term economic success of the so-called green revolution, but has caused a long-term degradation of the environment (Hough 1998). These chemicals have entered people's lives and surroundings. As is discussed in greater detail by Abdullah and Sinnakkannu (this volume), there are several ways in which the environment can be adversely exposed to these pollutants.

- The pesticides used for agricultural purposes are quite persistent and resist natural degradation processes. As a result, they remain in the soil in agricultural areas and then leach away with water to rivers and coastal areas as well as underground aquifers.
- Polluted run-off from agricultural lands adversely impacts on ecosystems that it comes in contact with. As an example, fish populations in rivers and coastal mangroves often have high levels of pesticides.
- Food produced through these chemical technologies typically has residual levels of pesticides and insecticides.
- Improper use of these technologies has been linked to accidental exposure of farmers to pesticides. This problem is particularly acute in the East Asian region, where the level of education is generally low and programmes for training pesticide users are not particularly successful.

Another aspect of the green revolution has been extensive monocropping. The enhanced use of fertilizers also allows closer spacing of plants and multiple harvests. The loss of natural biodiversity means that plants which may otherwise provide resistance to pests and enrich the soil are missing. Researchers claim that agricultural systems containing a range of biodiversity, both plants and animal life, have a built-in protective mechanism. This means that the existing systems have greater vulnerability to pests and plant diseases and suffer from an accelerated level of soil quality degradation. Against this backdrop of the challenges to maintain food security and efforts to reduce adverse environmental impacts, several questions arise that force a rethink of the approach to these problems. For example, one may ask the following questions.

- How can we maintain food production at a long-term sustainable level without adversely impacting on the environment and critical ecosystems?
- What technologies and methods are successful in achieving this food

security in different parts of the world? How can we adapt these technologies, such as integrated pest management, to local conditions in East Asia?
- What role can the general public play in guiding the policies governing food security? And what role can NGOs play in this respect?
- How can we increase public awareness on these issues, particularly with access to multimedia, the internet, and global communication resources?

East Asian NGOs and environmental governance

Effective non-governmental organizations (NGOs) can serve to strengthen civil society and community movements, in addition to focusing on relatively narrower environmental issues. It is worth recalling that in Europe, the USA, and Japan, it was only as a result of well-organized citizen action and protest over many years that governments began to act on the environmental problems created by industries. The challenges to the NGOs operating in the South are in fact much more difficult due to repressive governmental policies and lack of financial and human resources (Parnwell and Bryant 1996). Nevertheless, NGO groups are typically in a position where they can serve as educators and mediators between public and governments.

The fact is obvious that NGOs in East Asia play a significant role in environmental conservation and governance. It may be argued that this role has been gradually increasing over the past two decades because of increasing public awareness of issues, a wealth of available scientific information, and access to global information resources via the internet. The most salient impact of NGOs has been on the general public in terms of raising awareness of environmental issues and motivating citizen action against environmental degradation (Parnwell and Bryant 1996). This, no doubt, has a trickle-down impact on environmental policy formulation. It is also important to keep in mind that the nature of activities and effectiveness of NGOs in the five countries studied, and indeed the East Asian region in general, varies considerably. This is reflective of the cultural diversity and different political scenarios and levels of economic/industrial development.

Despite the country-to-country variability, it may be worthwhile to look in an aggregated manner at three major categories of NGOs that deal with environmental issues. A summary of the activities and their potential effectiveness is succinctly presented in Table 10.4; the reader is pointed to the first three sections of this book to find out the role played by NGOs in the three sectors studied.

Table 10.4 Activities and effectiveness of NGOs in environmental governance

NGO category*	Type of activities	Overall effectiveness
Environmental issues *Government and public*	Raising public awareness and pressure on polluters through conducting campaigns, surveys, and research; assessing policies and projects, and proposing alternatives; monitoring the environment, particularly air pollution.	Important role in educating and informing the public; positive impact in the formulation of national environmental plans and policies; mobilizing interest in and organizing major national conferences on the environment.
Farmer groups *Farmers, plantation workers, smallholders, and government*	Articulating problems related to farmers' livelihoods and living conditions; consumer education through discussions, house-to-house counselling, and so on; dissemination of publications; conducting scientific and social research; organizing conferences and training activities.	Improving knowledge of farmers; positive impact on increasing awareness among farmer communities; improvement in the national laws governing the agricultural sector.
Industry-oriented action *Pesticide manufacturing industries and government*	Organizing academic activity and training courses; carrying out environmental investigations and assessments; presenting recommendations for governments' decisions in formulating policy, strategy, and plans for pesticides; studying the strategy for the development of pesticide science and technology; organizing symposia and seminars.	Positive in the role of the bridge between governments and industries; an important role in promoting the development of pesticide science and industries and domestic and international exchange in the pesticide industry.

* The target audiences for each NGO group are shown in *italics*.

Recommendations for environmental governance in East Asia

Institutional aspects of environmental governance

The East Asian countries rely primarily on a command-and-control approach, where effluent and emissions standards are used as regulatory tools (Paoletto and Termorshuizen, Chapter 9). Effective institutions for implementation, monitoring, and compliance are typically not present. Generally speaking, a rigid command-and-control approach leads to high compliance costs and widespread under-compliance. To take the argument a step further, it may be proposed that it is difficult to implement even a strict command-and-control regulation in East Asia. This, in part, is driven by the lack of enforcement capacity as well as insufficient human and financial resources for large-scale investment in environmental protection. It is not uncommon to find a situation where environmental regulators ostensibly have the necessary detailed information, yet the polluters have the incentives to misrepresent their true costs without being held accountable. Therefore, it may be argued that unless regulators have the adequate capacity for monitoring and enforcement no effective environmental governance can be implemented.

As an example of the command-and-control approach, the environmental regulation system in China has not been effective because the authorities have set effluent disposal costs below the marginal cost of reducing pollution. Therefore, it is cheaper for factories to pay the disposal cost instead of reducing emissions. A better example is the Malaysian situation, where the water effluent standards are set as concentration. In principle, such an arrangement potentially provides the opportunity for dilution rather than real reduction. However, there has been effective implementation in Malaysia due to consistent and effective monitoring programmes developed in the country (Ahmad and Ali, Chapter 5).

Several Asian countries have begun to introduce market mechanisms for environmental management on an experimental basis. Their successes and failures highlight some of the limitations that must be taken into account in the East Asian context. Most importantly, fees and permits for effluents require effective monitoring. This is feasible for medium and large stationary sources, but is less so for small-scale scattered pollution sources. It should be noted that flexible environmental regulations are not appropriate in cases where even small quantities of pollutants can cause significant damage, such as toxic or hazardous waste.

Economic instruments and market mechanisms, however, permit cost saving and can lead to sustainable growth. A policy model based on a mix of command and control and market-based mechanisms linked to effective governmental management is showing positive results in countries like Malaysia, and is characterized by the role of government as a facilitator rather than provider; by a prominent role played by the private sector and civil society; and by pricing reform on environmental goods and services (UN ESCAP 2000).

- This model appears to have the greatest potential for developing countries of the East Asian region which are lacking financial resources.
- The market orientation of the industrial sector offers an opportunity to use market-based pollution controls more effectively.
- A flexible anti-pollution approach would use economic instruments to substitute for or complement regulatory standards.

Such a flexible approach can reduce the overall costs. For example, based on a sample of 260 enterprises in Beijing and Tianjin with multiple water pollution sources, Dasgupta *et al.* (1996) found that an emission charge that would achieve the current abatement rate for each pollutant would reduce abatement costs from $47 million to $13 million per year, a saving of $34 million from this group of enterprises alone, or a 70 per cent reduction from the cost of the command-and-control system. As another example, organizations with ISO 14001 certification can receive priority in bidding for governmental contracts. This can eventually lead to "greening" of the market-place.

Reducing the environmental impacts of industrialization

Environmental protection policy should be introduced with a view to the long-term sustainability of the economy rather than any short-term benefits. Failure to introduce environmental protection measures would in the longer term imply a loss of income as a result of the depletion of the natural resource base. For example, implementation of the polluter pays principle can have dramatic effects on how industries view their waste effluents and treatment processes. Similarly, eco-efficiency is the way forward for corporations which want to achieve success while taking account of environment and development issues (Schmidheiny 1992). This concept involves the production of goods and services while reducing resource consumption and pollution.

The industrial sector should emphasize development of technological fixes for cleaning and greening of industries. Transfer of technologies from developed countries can be a key component of this process. Obviously, technological developments have the potential to bypass many of the constraints which were once seen as threats to economic growth.

Reliance on information management and dissemination

Information generation, management, and dissemination are critical components of environmental governance. The East Asian nations can capitalize on the recent advances in information technology to their best advantage. A number of excellent examples are already present. The Chinese registry of potentially toxic chemicals, as reported by Hao and Yeru in Chapter 3, is a clear example of increased levels of transparency in information dissemination. Similarly, Tabucanon (Chapter 6) reports the availability of state of the environment and water quality data through a governmental website in Thailand. Japan has also established similar dissemination mechanisms through the internet (Yamauchi, Chapter 8). These trends will eventually take environmental governance to a higher level where civil society is well informed about issues and closely involved in the process. The role of NGOs in serving as catalysers of such developments cannot be overemphasized.

REFERENCES

ADB. 2000. *Asian Development Outlook 2000*. Hong Kong: Oxford University Press.

Dasgupta, S., M. Huq, D. Wheeler, and C. Zhang. 1996. *Water Pollution Abatement by Chinese Industry: Cost Estimates and Policy Implications*. Washington, DC: World Bank.

Hempel, L. C. 1996. *Environmental Governance: The Global Challenge*. Washington, DC: Island Press.

Hough, P. 1998. *The Global Politics of Pesticides: Forging Consensus from Conflicting Interests*. London: Earthscan Publications.

Mitchell, J. K. 1996. *The Long Road to Recovery: Community Responses to Industrial Disasters*. Tokyo: UNU Press.

Parnwell, M. J. G. and R. L. Bryant (eds). 1996. *Environmental Change in South-East Asia: People, Politics and Sustainable Development*. London: Routledge.

Satterthwaite, D., J. Hardoy, and D. Mitlin. 1992. *Environmental Problems in Third World Cities*. London: Earthscan Publications.

Schmidheiny, S. 1992. *Changing Course : A Global Business Perspective on Development and the Environment*. Cambridge, MA: MIT Press.

UN ESCAP. 1997. "Agricultural biomass energy technologies for sustainable rural development", paper presented to Expert Group Meeting on Utilization of Agricultural Biomass as an Energy Source, United Nations, New York, 16–19 July.

UN ESCAP. 2000. *State of the Environment in Asia and the Pacific – 2000*. Bangkok: United Nations Economic and Social Commission for Asia and the Pacific.

UNIDO. 2000. *International Yearbook of Industrial Statistics 2000*. United Nations Industrial Development Organization. Williston: Edward Elgar Publishing.

Acronyms

ACIAR	Australian Centre for International Agricultural Research
ADB	Asian Development Bank
ADI	average daily intake
AOAC	Association of Analytical Communities
APEC	Asia Pacific Economic Cooperation
ASEM	Asia-Europe Meeting
BBS	bulletin board system
BHC	benzene hexachloride
BLEPC	Basic Law for Environmental Pollution Control (Japan)
BMA	Bangkok Metropolitan Area
BOD	biological oxygen demand
BPMC	butylphenyl methylcarbamate
CAP	Consumers Association of Penang
CCPA	Canadian Chemical Producers Association
CCPR	Codex Committee for Pesticide Residues (FAO)
CDC	Center for Disease Control and Prevention (USA)
CFC	chlorofluorocarbon
CIF	cost, insurance, and freight
CO	carbon monoxide
COD	chemical oxygen demand
CMA	Chemical Manufacturers Association (USA)
CSPS	Chinese Society of Pesticide Science
DDE	1,1-dichloro-2,2-bis(chlorophenyl) ethylene
DDT	dichloro diphenyl trichloroethane
DEQP	Department of Environmental Quality Promotion (Thailand)

DID	Department of Irrigation and Drainage (Malaysia)
DIW	Department of Industrial Work (Thailand)
DMR	Department of Mineral Resources (Thailand)
DO	dissolved oxygen
DOA	Department of Agriculture (Malaysia, Thailand)
DOAE	Department of Agricultural Extension (Thailand)
DOC	Department of Chemistry (Malaysia)
DOE	Department of Environment (Malaysia)
EAJ	Environment Agency of Japan
EC	European Community
EDDP	ediphenphos
EHS	environmental health and safety
EIA	environmental impact assessment
EIC	Environmental Information Centre (Japan)
EKMA	empirical kinetic modelling approach
EMAS	European Eco-management Auditing Scheme
EMS	environmental management system
EPA	Environmental Protection Agency (USA)
EPN	o-ethyl o-4-nitrophenyl phenylphosphonothioate
EQA	Environmental Quality Act (Malaysia)
EQR	Environmental Quality Regulation (Malaysia)
EQS	environmental quality standard (Japan)
ERTC	Environmental Research and Training Center (Thailand)
EU	European Union
FAO	Food and Agricultural Organization
FDI	foreign direct investment
FTCSC	Federal Territory Counselling and Service Centre (Malaysia)
GC/MS	gas chromatograph and mass spectrometer
GDP	gross domestic product
GEACS	Greater East Asia Co-prosperity Sphere
GEIC	Global Environmental Information Centre (Japan)
GIFAP	International Group of National Associations of Manufacturers of Agrochemical Products
GKU	Green Korea United
GNP	gross national product
GSD	Geological Survey Department (Malaysia)
GWP	Global Water Partnership
HAP	hazardous air pollutant
HC	hydrocarbon
HC1	hydrochloric acid
HCB	hexachlorobenzene
HCH	hexachlorocyclohexane
ICAMA	Institute for the Control of Agrochemicals of the Ministry of Agriculture (China)
IEAT	Industrial Estate Authority of Thailand
INGO	international non-governmental organization

IPM	integrated pest management
IRPTC	International Register of Potentially Toxic Chemicals
IT	information technology
JARING	joint advance research integrated networking
JEC	Japan Environment Corporation
KADA	Kemubu Agricultural Development Authority (Malaysia)
KFEM	Korean Federation for Environmental Movement
LNG	liquefied natural gas
LPG	liquefied petroleum gas
MACA	Malaysian Agricultural Chemical Association
MADA	Muda Agricultural Development Authority (Malaysia)
MARDI	Malaysian Agricultural Research and Development Institute
MD	Mines Department (Malaysia)
MHW	Ministry of Health and Welfare (Japan)
MITI	Ministry of International Trade and Industry (Japan)
MOAC	Ministry of Agriculture and Cooperation (Thailand)
MOE	Ministry of Environment (Korea)
MOFA	Ministry of Foreign Affairs (Japan)
MOH	Ministry of Health (Malaysia)
MOI	Ministry of Industry (Thailand)
MOSTE	Ministry of Science, Technology, and Environment (Thailand)
MOT	Ministry of Transport (Japan)
MRL	maximum residue limit
MSMA	monosodium methylarsonate
MWA	Metropolitan Waterworks Authority (Thailand)
NAP	National Agricultural Policy (Malaysia)
NEB	National Environmental Board (Thailand)
NEPA	National Environmental Protection Agency (Thailand)
NESDB	National Economic and Social Development Board (Thailand)
NGO	non-governmental organization
NIER	National Institute of Environmental Research
NIES	National Institute for Environmental Studies (Japan)
NO	nitrogen monoxide
NO_x	nitrogen oxides
NRPTC	National Register of Potentially Toxic Chemicals
NRW	non-revenue water
NWC	National Water Council (Malaysia)
NWP	national water policy (Malaysia)
NWRC	National Water Resources Council (Malaysia)
OBD	on-board diagnosis
OC	organochlorine
OECD	Organization for Economic Cooperation and Development
OEPP	Office of Environmental Policy and Plan (Thailand)
OP	organophosphate
O_x	oxidant
PAH	polycyclic aromatic hydrocarbons

PAN	peroxyacetyl nitrate
PAN	Pesticide Action Network
PCB	polychlorinate biphenyl
PCD	Pollution Control Department (Thailand)
PEPB	Environmental Protection Bureau (China)
PER	Project on Ecological Recovery (Thailand)
PIC	prior informed consent
PIRG	Public Interest Research Group (USA)
POP	persistent organic pollutant
ppb	parts per billion
ppbC	parts per billion of carbon
ppbV	parts per billion by volume
PPP	polluter pays principle
PPSD	Plan Protection Service Division (Thailand)
ppt	parts per trillion
pptv	parts per trillion by volume
PRPTC	Provincial Register of Potentially Toxic Chemicals
PRTR	Pollutant Release and Transfer Register
PWA	Provincial Waterworks Authority (Thailand)
PWD	Public Work Department (Malaysia)
R&D	research and development
ROS	Registrar of Society (Malaysia)
RM	Malaysian ringgit
SAM	Sahabat Alam Malaysia (Friends of the Earth Malaysia)
SBPC	Small Businesses Promotion Council (Japan)
SCONTE	Society for the Conservation of National Treasures and Environment (Thailand)
SO_2	sulphur dioxide
SO_x	sulphur oxides
SPM	suspended particulate matter
SPWD	State Public Work Department (Malaysia)
SS	suspended solids
TCB	total coliform bacteria
TECDA	Thai Environmental and Community Development Association
TIE	trade-industry-energy
TSP	total suspended particulate
UNCED	United Nations Conference on Environment and Development
UNDP	United Nations Development Programme
UNECE	United Nations Economic Commission for Europe
UNEP	United Nations Environment Programme
UN ESCAP	United Nations Economic and Social Commission for Asian and the Pacific
UNFCCC	United Nations Framework Convention on Climate Change
UNU	United Nations University
VOC	volatile organic compounds
WHO	World Health Organization

WMA	Wastewater Management Agency (Thailand)
WPC	water pollution control
WSC	watershed classification
WWF	Worldwide Fund for Nature
YWCA	Young Women's Christian Association

Contributors

Dr Abdul Rani bin Abdullah, Alam Sekitar Malaysia Sdn. Bhd., No 19 Jln Astaka U8/84, Bukit Jelutong Business & Technology Centre, 40150 Shah Alam, Selangor, Malaysia.
Tel: 603-78454566
Fax: 603-78453566
E-mail: rani@enviromalaysia.com.my

Dr. Zafar Adeel, Assistant Director (Program Development), United Nations University, International Network on Water, Environment and Health, UNU/INWEH, McMaster University, Downtown Centre, 1st Floor, Hamilton, Ontario, Canada L8S 4L8.
Tel: +1 (905) 525-9140 extension 23082
Fax: +1 (905) 529-4261
E-mail: adeelz@inweh.unu.edu
UNU/INWEH at www.inweh.unu.edu/unuinweh

Dr. Abdul Rashid Ahmad, Associate Professor, Department of Geology, Faculty of Science, University of Malaya, 50603 Kuala Lumpur, Malaysia.
Tel: 60379674156
Fax: 60379675149
Email: abrashid@um.edu.my

Dr. Hasnah Ali, Associate Professor, Head, Department of Economic development, Faculty of Economics, Universiti Kebangsaan Malaysia (National University of Malaysia), 43600 Bangi, Malaysia.
Tel: 60389215778 60389213741
Fax: 60389213260
Email: hasnah@pkrisc.cc.ukm.my

Dr. Quan Hao, Former Director General, China-Japan Friendship Center for Environmental Protection, State Environmental Protection Administration, No. 1

Yuhui Nanlu, Chaoyang District, Beijing 100029, China.

Dr. Meehye Lee, Associate Professor, Department of Earth & Environmental Sciences, Korea University Anam-dong, Sungbuk-gu, Seoul 136-701, South Korea.
Tel: +82-2-3290-3178
Fax: +82-2-3290-3189
Email: meehye@korea.ac.kr

Ms. Naori Nakamoto (Miyazawa), International Program Officer, Environment and Sustainable Development Specialist, International Development Center of Japan, East Timor Office, P.O. Box 184, Dili, East Timor.
Tel: +670 390 322774
Fax: +670 390 312160
Email: miyazawa.n@idcj.or.jp
www.idcj.or.jp

Mr. Glen Paoletto, Capacity Building Programme, Institute for Global Environmental Strategies (IGES), 1560-39, Kamiyamaguchi, Hyama, Miura-gun, Kanagawa, Japan 240-0198.

Dr. Saraswathy Sinnakkanu, Chemistry Lecturer, International Education Centre, University Technology MARA, Section 17 Campus, 40200 Shah Alam, Selangor, Malaysia.
Telephone office: 603-55482430
Fax: 603-55414773
E-mail: sttan@acmar.po.my

Dr. Monthip Sriratana Tabucanon, Director-General, Department of Environmental Quality Promotion, Ministry of Science and Technology, Bangkok, Thailand.
Email: monthip@deqp.go.th

Ms. Cindy Termorshuizen, Institute for Global Environmental Strategies (IGES), 1560-39, Kamiyamaguchi, Hyama, Kanagawa, Japan 240-0198.
Tel: +81-468-55-3700
Fax: +81-468-55-3709
URL: http://cb.iges.or.jp

Ms. Makiko Yamauchi, School of Oriental & African Studies, Thornhaugh Street, Russell Square, London WC1H0XG, UK.

Dr. Huang Yeru, Associate Professor, China-Japan Friendship Center for Environmental Protection, State Environmental Protection Administration, Beijing 100029, China.
Tel: +86-10-64947722
Fax: +86-10-64966344
Email: quanhao@public3.bta.net.cn

Index

Catalogue Request

Name: ______________________________

Address: ______________________________

____

Tel: ______________________________

Fax: ______________________________

E-mail: ______________________________

To receive a catalogue of UNU Press publications kindly photocopy this form and send or fax it back to us with your details. You can also e-mail us this information. Please put "Mailing List" in the subject line.

53-70, Jingumae 5-chome
Shibuya-ku, Tokyo 150-8925, Japan
Tel: +81-3-3499-2811 Fax: +81-3-3406-7345
E-mail: sales@hq.unu.edu http://www.unu.edu